NONPARAMETRIC STATISTICS FOR SOCIAL AND BEHAVIORAL SCIENCES

NONPARAMETRIC STATISTICS FOR SOCIAL AND BEHAVIORAL SCIENCES

M. KRASKA-MILLER

Auburn University
Alabama, USA

CRC Press
Taylor & Francis Group
Boca Raton London New York

CRC Press is an imprint of the
Taylor & Francis Group, an **informa** business

A CHAPMAN & HALL BOOK

CRC Press
Taylor & Francis Group
6000 Broken Sound Parkway NW, Suite 300
Boca Raton, FL 33487-2742

First issued in paperback 2019

© 2014 by Taylor & Francis Group, LLC
CRC Press is an imprint of Taylor & Francis Group, an Informa business

No claim to original U.S. Government works

ISBN-13: 978-1-4665-0760-9 (hbk)
ISBN-13: 978-0-367-37910-0 (pbk)

Library of Congress Cataloging-in-Publication Data

Kraska-Miller, M. (Marie)
   Nonparametric statistics for social and behavioral sciences / M. Kraska-Miller, Auburn University, Alabama, USA.
       pages cm
   Includes bibliographical references and index.
   ISBN 978-1-4665-0760-9 (alk. paper)
   1. Social sciences--Statistical methods. 2. Psychology--Statistical methods. I. Title.

HA29.K798 2014
519.5--dc23   2013026481

Visit the Taylor & Francis Web site at
http://www.taylorandfrancis.com

and the CRC Press Web site at
http://www.crcpress.com

Contents

Preface

This textbook is written for students, instructors, and practitioners who seek a practical introduction to nonparametric statistics. *Nonparametric Statistics for Behavioral and Social Sciences* is an appropriate textbook for a complete course in nonparametric statistics for upper-class undergraduate students and graduate students in the behavioral and social sciences fields of study. Students in beginning statistics courses that include only a unit on nonparametric statistics will find this book a useful resource as well. Also, students and researchers in disciplines other than social and behavioral sciences may find it a useful reference and guide for common nonparametric statistical procedures. For example, students in veterinary medicine, pharmacy, rehabilitation, agriculture, and business may find this book useful. This statement is made with confidence because practitioners of these disciplines often conduct research using small data sets that do not meet the requirements of parametric statistics.

Although the primary use of this book is for teaching, it is also appropriate for use as a research reference by school administrators, counselors, and teachers. In addition to serving novice researchers, this textbook is intended for those who already have some research and/or data analysis experience. Local professional school personnel may be interested in such a book for conducting research and also as a reference in reviewing the literature.

After years of teaching statistics without a textbook specifically dedicated to nonparametric procedures in behavioral and social sciences, the author is convinced that such a book needed to be developed. *Nonparametric Statistics for Behavioral and Social Sciences* emphasizes practical applications of theory in various fields of study. It covers the concepts, principles, and methods used in performing many nonparametric procedures. It is the only nonparametric book on the market written specifically for students in the behavioral and social sciences and demonstrates the most common nonparametric procedures using software by SPSS, Inc., an IBM company.[*]

In talking with many of my students, I learned that they welcome a textbook that shows nonparametric data analysis procedures in a research context. In other words, this book shows the connection between research and data analysis without burdening students with more technicalities than necessary. This is a user-friendly textbook that uses a nonmathematical approach to nonparametric procedures with minimal coverage of formulas.

In writing this book, the author sought to achieve the following overall goals: (1) provide instruction in the applications of nonparametric statistics to real-life research problems, (2) present information from a conceptual rather

---

[*] IBM acquired SPSS in October 2009.

than a mathematical view, (3) ensure clarity of procedures, and (4) present information in user-friendly language and a readable format. The book is not written as a "cookbook," nor is it intended to be. However, certain procedural steps are necessary to guide readers' use of the software. Such procedures are presented at the points where they are required. Screen shots of the steps of procedures and the resulting statistical output along with interpretations are included to aid understanding and learning. The many charts, graphs, and tables included help students visualize the data and the results. A hands-on pedagogical approach helps enhance learning as do the exercises following each chapter.

Statistical formulas and explanations are given as needed for each procedure. The formulas are for illustration purposes only, and it is unnecessary for readers to have quantitative backgrounds. Prerequisite courses in mathematics and statistics are not required for effective use of this book.

Its specific goals are to (1) discuss a conceptual framework for each statistical procedure, (2) present examples of appropriate research problems, associated research questions, and hypotheses that precede each statistical procedure presented, (3) detail IBM-SPSS paths to conducting various analyses, and (4) explain interpretations of statistical results and conclusions of the research.

Explanations of appropriate procedures for research designs are given. The logic and purpose of specific procedures are presented in nontechnical terms so that students can learn the concepts. All technical terms required to understand procedures are explained. In addition, databases and data input methods using SPSS for specific procedures are explained and illustrated, and examples of procedures are included. Each chapter includes explanations, illustrations, examples, interpretations of results, and student exercises for each statistical procedure presented. SPSS screen shots show steps of procedures. Chapter 1 provides an introduction to research and explains various types of research designs and steps in the research process. This is an important chapter because it provides a basis for a student to plan a research study. Statistical analysis without empirical and/or theoretical bases is meaningless. This first chapter is intended to guide students in developing plans for a research study.

Chapter 2 presents a discussion of the differences between nonparametric and parametric statistical methods. Nonparametric statistical procedures are important to analyze data that do not meet the requirements of parametric statistics.

Chapters 3 through 8 discuss a related group of nonparametric statistical procedures. Chapter 3 covers methods of analyzing data for association and agreement. Chapter 4 discusses analyses of two independent samples, and Chapter 5 presents methods for analyzing multiple independent samples. Analysis of two dependent samples is presented in Chapter 6, and methods for testing multiple related samples are presented in Chapter 7. Chapter 8 focuses on procedures for analyzing single samples.

High-speed computers and advanced statistical software make data analysis available and accessible for those who need or desire to conduct research. The SPSS statistical software program is demonstrated in a series of practical applications throughout this book. The focus of this book is on nonparametric statistical procedures that require understanding and using statistical software critical to conducting analyses. SPSS is a menu-driven program for students in behavioral and social sciences disciplines such as education, sociology, psychology, consumer affairs, and others.

Sound research designs, appropriate statistical analyses, and accurate interpretations of results are required to advance any discipline. This textbook is written for such purposes. It is the author's hope that this textbook proves to be an effective and efficient tool for those who choose to use it.

Acknowledgments

Any undertaking as huge as writing a textbook requires the support of many people. Without a supportive editor and editorial staff, such a project cannot come to fruition. I acknowledge David Grubbs, my editor, who supported and guided me throughout this project. David gave me the latitude I needed to develop a statistics textbook for students, educators, and others in the social and behavioral sciences. He was always willing to listen to my ideas for content and make suggestions. I am grateful that David was my editor for this project.

Also, I express my appreciation for the work that Marsha Pronin, my project coordinator at Taylor & Francis, devoted to this project. Marsha provided an opportunity for me to participate in the production of the manuscript and sought my ideas on design and appearance of the final product. She gave careful attention to every detail as the manuscript went through the production stages, and she kept me informed throughout the process.

In addition, I acknowledge Dr. Gregory Petty, biostatistics professor in the Department of Public Health, University of Tennessee, who reviewed the manuscript and provided valuable feedback that resulted in a stronger and more comprehensive textbook. I appreciate his willingness to share his time and expertise.

I am also indebted to the students in my nonparametric course during the Spring 2013 semester who took extra care in reading the manuscript. Their questions and suggestions helped me to make the content more clear, complete, and practical. I owe a debt of gratitude to each and every one of them for assisting me in field testing this book. Their sincerity and patience in studying the material exceeded my expectations.

I acknowledge the patience, support, and encouragement of Dr. W. R. Miller, my husband and mentor. I am truly blessed that throughout this project, he was always willing to make adjustments in his schedule to accommodate my deadlines.

Finally, I express my appreciation to all of the staff at Taylor & Francis Group who were involved in the project development phase to help make this manuscript come to life.

M. Kraska-Miller

1

Introduction to Research in Social and Behavioral Sciences

Professionals at all levels in social and behavioral fields of study benefit from research. For example, curricula changes, organizational policies, approaches to sociological inquiries, and instructional practices are predicated on the results of sound research. Research allows individuals to make objective and informed decisions so that they may use scarce resources wisely.

Results from research studies allow decision makers to base their decisions and subsequent actions on a solid foundation. Local, district, state, and national accountability require objective and measurable outcomes. Data-based research is paramount to effective decision making. For example, school systems apply the results of research to decisions such as the number of school buses needed, amount of food necessary to feed students, priorities for building maintenance, and the number of classrooms and laboratories needed to serve students. While many may think that such decisions are automatic, all decision makers should have a basic working knowledge of research. In other words, decisions cannot be made effectively without supporting evidence, regardless of the simplicity of the questions.

The contributions of research to improvements in sociology, psychology, education, and other human sciences are numerous. For example, professional educators know a lot about teaching, learning, and achievement motivation; however, gaps remaining in the knowledge base warrant further investigation. Thus, a major role of research is to close the gaps between existing knowledge and desired knowledge. Research is also important to expand already existing knowledge bases.

Not all research is aimed at discovering new information. Sometimes studies are conducted to help validate previous research findings. Whatever the specific goal of research, the focus should be on sound research designs and practices. Individuals who read research reports position themselves to be more effective and efficient practitioners, consumers of research, and active researchers because they are able to evaluate and apply research results to their professional practices. Being an informed consumer of research is as important as conducting research in terms of bringing overall improvements to a specific field of study.

Before the research process can be applied, one must acknowledge that a problem exists. After the problem is acknowledged, the next step is to define and describe it. Defining a research problem is crucial; otherwise researchers may waste valuable time and resources investigating areas that do not directly address the problem of interest. Research problems should be anchored in a theoretical or empirical basis. This means that the researcher should conduct a study to support or refute an existing theory, confirm some observations, and/or generate new evidence. Whatever the basis for research, the researcher's first task is to establish a sound basis as a starting point before applying the basic principles of research.

This chapter provides the context within which subsequent chapters describe the various statistical procedures that are applied to data resulting from the research process. It begins with an introduction to the basic principles of research. Next, information related to planning is presented. Various types of research designs are discussed, along with examples that illustrate applications. Also, information about sampling procedures and different kinds of samples is included. In addition, this chapter presents a discussion of the validity and reliability of measurement instruments. A section on data analysis explains the difference between descriptive and inferential statistics. Finally, the steps in the research process are described. The chapter ends with a summary of the concepts and principles discussed. Student exercises are included to reinforce learning and to provide opportunities to apply the information presented.

1.1 Basic Principles of Research

Research involves a study or investigation of the issues, conditions, problems, and concerns related to a field of study or discipline. The research process involves the systematic collection of data that are analyzed and interpreted so that logical conclusions may be drawn. One may think of data in a research study as the raw material from which evidence is generated. It is the interpretation of the evidence that can provide answers to questions and solutions to problems. Sometimes data may be collected simply to create new knowledge that adds to an already existing knowledge base.

Most research studies in social and behavioral sciences seek new information that is beneficial to practitioners, policy makers, and planners. Therefore, the primary purpose of research in the social and behavioral sciences is to increase the body of knowledge related to various aspects of a discipline or field of study, such as education, psychology, kinesiology, management, or sociology. Some studies use existing numerical data and information to bring about a new perspective that makes it possible for researchers to gain greater insights or learn new applications.

A systematic investigation requires that researchers specify a well-defined purpose for a study, the objectives to be achieved, the questions to be answered, or hypotheses to be tested. Special tools and instruments needed to collect the necessary data are selected, and the procedures for data collection are determined. After potential participants are identified, data are collected and analyzed. Finally, researchers interpret the results of the analysis and draw appropriate conclusions. Upon completion of a study, a report should be prepared to describe the findings and appropriate conclusions to be drawn from the investigation.

Research requires scientific thinking. Scientific thinking is an orderly process whereby conclusions are based on factual evidence, and logic is used to demonstrate relationships among related ideas. Scientific thinking involves movement of thought between inductive and deductive reasoning. Researchers may use inductive reasoning in the process of taking partially known, fragmentary evidence gained in a variety of ways and moving toward broad generalizations. Deductive reasoning is employed when researchers use systematic procedures to test a given conclusion or principle. The results may lead to confirmation, acceptance, or modification of the principle or generalization.

The application of scientific thinking to research requires systematic inquiry, factual evidence, and analysis of the evidence to obtain accurate conclusions. However, as important as the research process is, factors of even greater importance are the initial identification and definition of the problem to be investigated. In fact, without a clearly defined statement of the problem and logically derived hypotheses, a researcher has an inadequate basis for determining the evidence, i.e., data needed for the investigation. Although the investigator's thoughts may move back and forth among the steps in the research process, there is a logical process for applying the scientific method to research.

Research requires the application of scientific procedures. The acquisition of new or reformulated knowledge is the goal of a research project. If such a body of knowledge is to be relied upon to guide professional practice, it must be acquired through the application of scientific methods. For research procedures to qualify as scientific endeavors, the characteristics listed in Table 1.1 should be present.

TABLE 1.1

Characteristics of Scientific Studies

1. Research designs are based on structured, rigorous, and ethical procedures.
2. Research results are based on objective evidence.
3. Scientific procedures permit replication of a study and verification of results.
4. Research questions and hypotheses are stated to guide a study.
5. Hypotheses to be tested are stated in the null form, tested empirically, and lack emotion-laden language.
6. A scientific approach controls for systematic bias in study design.
7. Science uses valid and reliable measuring tools.
8. Systematic studies involve scientific thinking.

Professionals in the behavioral and social sciences are both producers and consumers of research. Through knowledge and experience, researchers must become skillful in detecting inappropriate measurement tools, faulty data collection procedures, and applications of inappropriate statistical procedures. Errors in any of these critical areas lead to inaccurate and erroneous conclusions. Such conclusions are useless and possibly misleading to practitioners and future researchers.

Researchers must be objective when reporting the results of their investigations. They should present all the data, not just the data that support their hypotheses. Reporting all outcomes, including those that refute the original hypotheses, is the mark of a capable and ethical researcher. Finally, researchers must prepare reports that provide sufficient procedural detail to permit replication and verification of their findings.

1.2 Planning a Study

Researchers in the developed nations have devised systematic procedures to be followed when studies of problems in the social and behavioral sciences are planned and conducted. The initial task faced by a researcher is to identify significant problems that can be studied by the application of systematic procedures. The procedures to be applied are as varied as the problems to be investigated.

A research design is a plan the researcher puts in place to follow throughout the conduct of an investigation. The design must be appropriate for the purpose of the study and the research hypotheses to be tested, if any. In other words, the design of a study provides direction for the study, just as a map does for a navigator. A plan for quantitative research should include specific components to assure that the study is accurate, meaningful, feasible, efficient, and objective.

Quantitative research designs vary from highly rigorous with randomization, controls, varying treatments or conditions, and manipulation of independent variables, to less rigorous designs using intact groups, no opportunity for manipulation of independent variables, and no treatments or controls. The most rigorous designs are classified as experimental and the less rigorous are known as nonexperimental designs. Both types of designs follow various structures and involve varying degrees of rigor.

The research process is nonlinear. All parts of a project are related, and a researcher must recognize that several components may need to be planned simultaneously. Section 1.6 presents the components individually for ease of understanding and learning. As a research study develops and progresses, the researcher may move back and forth among the components for each of the steps in the process. For example, the theoretical basis of a study may be improved if new information supports or fails to support the researcher's original hypothesis. Modifications in sample selection methods or sample size may become necessary if time lines or budgets change.

Every research plan should include key components that provide direction for a study. Some components, such as establishing a theoretical and/ or empirical basis for a study are required for all studies. Some components are included only as necessary and dictated by the aim of each specific study. Therefore, the researcher should take care in planning a study.

The preparation of a research plan or proposal is an essential step in the research process. Regardless of whether a study is historical, descriptive, or experimental, a proposal is a prerequisite. The research proposal is a systematic plan to achieve the purpose of the study.

1.3 Types of Research Designs

Approaches to research may be classified into several categories, depending on the problem and purpose of a study. The purpose of a study is the researcher's rationale for conducting it and represents the specific function that the researcher hopes the results will serve. The design of a study is based on decisions that the researcher makes about critical components of a study such as the objectives, questions to be answered, hypotheses to be tested, sampling, instrumentation, analysis, and conditions.

The type of research conducted is contingent upon the intended purpose, for example, building a theory or studying the variables that make immediate impacts in practical situations. In these cases, the research may be classified as basic or applied, respectively, and both approaches are discussed in the following sections.

1.3.1 Basic Research

Basic research is usually conducted over a certain time period for the purpose of verifying or modifying an existing theory or generating a new one. The results of basic research are corroborated with existing research on the same or similar topics to determine whether support of the earlier theories holds. In a more general sense, empirical evidence that contradicts the theory is not discovered.

In the behavioral and social sciences, basic research is conducted to increase the knowledge base and understanding of societal issues or human behaviors. For example, psychologists study achievement motivation for the purpose of verifying existing theories and creating new ones. The theories derived from basic research serve as bases for many applied studies in the behavioral and social sciences. While researchers conduct basic research to study theories such as achievement motivation, other researchers conduct applied research studies to ascertain whether a specific motivational technique will work in practice with a certain group of individuals.

1.3.2 Applied Research

The purpose of applied studies is to solve some immediate or impending problem. Like basic research, applied research is also based on empirical evidence. The difference between the two approaches is that basic research uses the evidence to support theories and applied research uses the evidence to enhance practice. The results of applied research are used to improve practices, policies, and programs. For example, counselors seek ways to improve counseling practices, school administrators develop fair and reasonable policies affecting students, and curriculum designers work to improve academic and nonacademic programs.

Another classification scheme for research is based on the design of a study. The design of a study is a component of basic and applied research efforts. This section covers only research designs that require statistical analyses of numerical data. Research designs such as historical studies and case and field studies do not require such statistical analyses and are not covered here.

General classifications of research designs applied to behavioral and social sciences are known commonly as experimental, quasi-experimental, or nonexperimental. Within each general classification of design, a variety of modifications or configurations may be chosen. The purpose, characteristics, advantages and disadvantages, and special considerations are presented for each of the designs.

1.3.3 Experimental Design

The purpose of experimental studies is to investigate cause–effect relationships between two variables or among more than two variables under certain conditions or treatment situations. Specifically, the purpose is to identify whether groups differ on a specific variable (called the dependent variable) based on their exposure or lack of exposure to a specific condition or treatment (called the independent variable). Subjects in the comparison groups should display similar characteristics, especially characteristics that may influence the outcome of the dependent variable. True experimental designs satisfy the general criteria of randomness, control, and manipulation.

Randomness refers to the random selection of subjects and the random assignment of subjects or groups to treatments or conditions. Randomization is one way to control for sample bias. Each participant in a target population should have the same opportunity to be selected. Randomness can be achieved by the use of a table of random numbers, a computerized random number generator, or simply by pulling names out of a hat. Random selection of subjects and their random assignment to treatments help minimize systematic errors by controlling variables other than those of interest that can influence results of the study.

Variables that can influence the outcomes of a study are called extraneous variables. They compromise the validity of the results of experimental investigations. Examples of extraneous variables are existing support systems for clients in counseling, students' prior knowledge of certain academic content, and years of experience in management. Randomness is not any haphazard method of selecting subjects or assigning treatment conditions. The random selection procedure should be conducted systematically, with careful attention given to all important aspects of the independent variables. In other words, simply selecting subjects randomly without consideration for important ways in which they differ invites extraneous variables (also called confounding variables) into a study.

Experimental control is the researcher's prerogative to manage the research situation or setting. True experimental studies require strict controls of all critical elements. Strict controls have led some researchers in the behavioral and social sciences to question whether a controlled setting is too much like a laboratory. While internal validity may be enhanced, the strict controls may interfere with the external validity of a study. For example, in a study of student achievements under two different curricula, we cannot attribute improved performance to a curriculum if student performance is the same under both curriculum conditions.

Examples of researcher control in true experimental studies are the start time for the study, conditions under which the study will be conducted, and the duration of the study. Control is closely related to manipulation of the independent variables. A major characteristic of true experimental design is the researcher's freedom to manipulate one or more independent or predictor variables that are hypothesized to affect the outcome variable. In other words, manipulation involves a researcher's decisions about the conditions for the study or treatments (independent variables).

Manipulation can involve the management of one or more independent variables. Each independent variable may involve one or more levels, such as counseling methods for individuals, large groups, or small groups. A researcher can direct the treatments and the conditions to determine which subjects receive specific treatments. Manipulation involves decisions such as the types of experimental treatments or numbers of groups in a study. True experimental designs are difficult to conduct in the behavioral and social sciences because human nature is variable. One cannot hold all possible variables constant when studying human behavior.

The rigorous nature of true experimental designs is a major advantage to researchers. The features of randomization, control, and manipulation help to assure an objective and systematic approach to establishing causal relationships or verifiable differences among groups. True experimental designs are easy to replicate because they follow systematic methods. In addition, conclusions can be verified readily with similar repeated studies.

Replication and verification are hallmarks of sound research. When carried out properly, experimental studies produce results that are observable, replicable, verifiable, and reliable.

As with all research, experimental designs also have disadvantages. One is the possibility that the sample is not representative of the population to which the researcher wants to generalize. In addition, careful attention should be given to assure that the variables being manipulated are indeed the predictors or causes that can affect the outcomes. Other disadvantages are experimenter bias, weak controls, measurement error, artificiality of the study conditions, and faulty interpretations of results. Particular care in the conduct of experimental studies should be taken to ensure internal and external validity of the results. The following example illustrates an experimental design.

1.3.3.1 Example: Experimental design

A researcher will conduct an experimental study to assess reading comprehension. The study involves three groups of children selected randomly from five pre-kindergarten sections. The children are assigned randomly to one of three groups: one group of students with a reading coach, one group utilizing picture stories, and a control group following a traditional teacher approach. After uniform periods of instruction under the three conditions, each group of children is administered a reading comprehension test, and their final scores are compared to determine differences among the groups.

1.3.4 Quasi-experimental Design

When all criteria necessary for experimental designs are not present, researchers may design studies to mimic experimental designs as closely as possible. If random assignment to a specific treatment group is not possible and random selection of subjects is impractical or unethical, then a study is classified as quasi-experimental; for example, research in education and sociology is conducted often on existing groups (classrooms, social clubs, work teams). By the same token, if subjects can be selected randomly but random assignment to conditions (patients with macular degeneration and those without) is impossible, then a design is classified as quasi-experimental. In other words, quasi-experimental designs meet only some (not all) of the strict criteria for experimental designs. Obviously, quasi-experimental designs are not as powerful as experimental designs because researchers have no way to isolate causes or control extraneous variables.

One advantage of quasi-experimental design is that researchers can approximate experimental designs, albeit with some constraints as mentioned in the preceding paragraph. Another advantage is that quasi-experimental designs allow for hypothesis testing, making it possible for researchers to rule out rival hypotheses. Quasi-experimental designs are desirable in applied situations if a researcher has limited control over some or all of the independent variables. When pure experimental research cannot be conducted, a quasi-experimental design is certainly a viable alternative. Many research studies in the social and behavioral sciences use quasi-experimental designs because the subjects cannot be selected randomly or because conditions and/or treatments cannot be assigned randomly.

A major disadvantage of quasi-experimental design is the inability of researchers to control for extraneous variables through randomization. In addition, the lack of control may compromise internal and external validity. To some extent, quasi-experimental designs pose some of the same challenges as pure experimental designs: assuring sample representativeness and avoiding researcher bias. The following example demonstrates a quasi-experimental study.

1.3.4.1 Example: Quasi-experimental design

A guidance counselor is interested in the career awareness and occupational knowledge of high school seniors enrolled in college preparatory programs versus general high school curricula. The counselor randomly selects 100 students enrolled in a college preparatory program and another 100 students enrolled in a general high school curriculum. This study has the element of random selection from a clearly identified population; however, random assignment to a condition (program) is impracticable. The counselor administers an inventory of career awareness questions and occupational knowledge. Scores for both groups are compared.

1.3.5 Nonexperimental Design

When neither random selection nor random assignment to treatments is possible, ethical, or desirable, a research design is called nonexperimental. Nonexperimental studies are a general class of several different kinds of research designs in which random selections of subjects and random assignments to treatments or conditions are unethical, unnecessary, impractical, or undesirable. However, nonexperimental designs can investigate impacts of intact groups on a dependent variable. Descriptive or status quo studies, correlation studies, causal comparative studies, and longitudinal studies are classified commonly as nonexperimental. They do not utilize treatment or control groups. Nonexperimental designs rely on tests, surveys, and interviews as data sources rather than data obtained by designing experiments and analyzing results. Several common nonexperimental designs are presented in the following paragraphs.

1.3.6 Descriptive Design

Descriptive studies, as the name implies, are designed to investigate a current state of affairs. These studies are also known as status studies because the focus is on the impacts of groups, objects, or phenomena on variables of interest. Randomization of subject selection may or may not be possible or desirable, depending upon the size of the population and the nature of the study.

1.3.6.1 Examples: Descriptive design

The following research problems are examples of descriptive studies.

1. A curriculum planner is interested in the current status of developmental reading programs within junior colleges of a given state or region. This type of research problem would address factors such as the number of reading laboratories at identified colleges and geographic locations, the kinds of equipment, number of staff members, number of students served, hours of operation, and related items. Use of any additional information that helps to provide a picture of the reading laboratory conditions is appropriate.

2. A sociologist may be interested in expectations of voters under 30 years of age. Answers to questions related to social concerns, such as the environment, education, and employment, may be sought.

3. Students enrolled in a biostatistics course may collect information to describe certain medical markers such as blood pressure and glucose levels of patients.

1.3.7 Correlational Design

Correlational studies focus on linear relationships between two or more variables for one or more groups. The variables are represented by scores or values such as test scores for mathematics, reading, language arts, or history. Scores may also be based on attitudes toward social issues or voter opinions of candidates. Any association between the variables should not be interpreted incorrectly to mean that one variable causes the other. In other words, the variables in a correlational study are not considered to be predictors or criterion variables.

A common statement in the research literature is that correlation does not mean causation. The Pearson product–moment correlation coefficient range is ±1. A coefficient of 1 indicates a perfect linear relationship regardless of the sign. On the other hand, a correlation coefficient of 0 means absolutely no linear relationship between variables. A negative coefficient means that as one variable increases, the other decreases; whereas a positive sign means that as one variable increases, the other variable increases also.

The strength of a relationship is indicated by the size of the coefficient. A conservative rule of thumb is that coefficients of 0.1, 0.3, and 0.5 indicate weak, moderate, and strong relationships, respectively. When the value of one variable increases and the value of the other increases or decreases, the relationship is said to be linear. Not all correlational studies yield linear relationships. Special statistical procedures are used to evaluate nonlinear relationships.

Correlational studies have the advantage of simplicity of design. One needs only to identify the variables of interest and conduct the analysis. Another advantage is versatility. A correlational design may serve as a starting point in analyzing complex relationships for differences among variables. In addition, interpretation of results is fairly straightforward. Correlation studies allow researchers to determine the extent of agreement between and among variables (strong, moderate, weak) and the direction of agreement (positive or negative).

One disadvantage of correlational studies is that a third (or more) of unidentified variables may account for the correlation between two variables of interest. Researchers cannot exercise control over confounding variables. Correlational studies are sometimes considered to yield unsophisticated solutions to complex problems.

1.3.7.1 Examples: Correlational design

The following examples illustrate correlational studies.

1. A psychologist is interested in the extent to which children who relate positively to animals also relate positively to their peers.
2. A sociology professor wants to investigate the relationship between students' views of social capital, defined as trust and community development, and their self efficacy.

1.3.8 Causal Comparative Design

Not all studies lend themselves to the strict criteria of true experimental designs or even the limited criteria of quasi-experimental designs. Causal comparative designs are opposite in structure and conduct from experimental designs. In experimental designs, researchers attempt to isolate causes (variables) so that they can examine effects. In causal comparative studies, researchers observe effects and then look retrospectively for the causes. Causal comparisons are also known as ex post facto (after-the-fact) studies. Observed effects, although they already occurred along with the causes, serve as the dependent variables. Causal comparative research is valuable for many studies in the social and behavioral sciences such as investigating patterns of behavior and achievement motivation by examining past school records of children. Sociologists can use causal comparative research to study election outcomes by examining age, socio-economic status, race, geographic region, religion, and other voter attributes.

An advantage of causal comparative studies is that researchers can investigate problems in which randomization, control, and manipulation are not possible, ethical, or desirable. Studies that are too costly or impractical to conduct as true experiments may be conducted with causal comparative methods.

An important disadvantage of causal comparison studies is the difficulty in ensuring that subjects are similar in all respects except the variables of interest. Second, the lack of control of independent variables can limit conclusions that researchers can draw from the findings. Researchers must work within the confines of the available data and exclude all variables outside those in the existing database. Researchers may be challenged to prove that the variables studied are indeed the causes of the observed effects. This difficulty is compounded by the fact that ex post facto studies provide no assurance that relationships uncovered in an analysis are not products of unknown or unanalyzed variables. Regardless of the shortcomings of causal comparative studies, existing databases provide a rich and ready source of information for researchers desiring to address real-life problems.

The following examples use medical situations to compare experimental and causal comparative designs.

Experimental design: Patients who are overweight are asked to participate in a study that requires them to follow a new diet or an exercise program for six months. A physician will check changes in patient weights after six months of participation.

Causal comparative design: Physicians try to identify reasons for patients' heart conditions by comparing patients' diets and lifestyles to individuals who do not have heart conditions.

1.3.9 Longitudinal Design

Longitudinal studies can be classified as quasi-experimental or nonexperimental, depending on the level of control and purpose of the study. Such studies collect data on the same variables at designated points in time to measure changes in dependent variables from one time point to another. Myriad problems in social and behavioral sciences, medicine, engineering, and business can be explored with longitudinal studies.

Longitudinal studies have the advantage that measures of the same variables from one time period to another can be compared. It may be something as simple as measuring the growth of children from one year to the next or a more complex measurement such as adjusting a dosage of medicine over time. While repeated measures over time can uncover relationships and solutions to problems, the time lag between measures can be problematic.

Disadvantages of longitudinal studies include the difficulty in controlling independent variables, especially when studies continue over months or years. Consequently, extraneous variables can confound the results. Depending upon the duration of the study, participant mortality or attrition (dropout) can threaten internal and external validity. Longitudinal studies, by their very nature, suggest that researchers make a trade-off between loss of statistical power or not conducting a study at all. The time commitment required of researchers and the resources required for analyses of repeated measures may not be feasible if external funding is limited.

1.3.9.1 Example: Longitudinal design

A marriage counselor follows newlywed couples for 12 years to evaluate her hypothesis that couples who have marital problems in the 7th and 11th years of marriage will file for a divorce in those years. The study includes only data from couples who are still married to their first spouses at the end of the 12 years. Couples who agree to participate in the study are encouraged to remain for the duration of the study. The counselor contacts the couples annually to verify their information. The counselor collects baseline data at year one to evaluate the openness and marital satisfaction of the couples. At years 3, 7, 9, and 11, the same questionnaires about openness and marital satisfaction are administered to the couples. Table 1.2 lists the types of research designs and their distinguishing features.

1.4 Sampling Procedures

The population from which information is to be gathered is frequently too large to be practical; therefore, a subset or sample of the population must be used. The manner in which the sample is selected is critical to a successful study. Several sampling procedures may be used. When researchers select samples, they must utilize a rationale that justifies the process used to select the participants, objects, or events to be investigated. Time spent studying a biased or inappropriate sample is time wasted. What might have been a successful project may be biased from the beginning by inadequate sampling. Consequently, a thoughtful and careful approach to sample selection supports the credibility and integrity of the final results. Each sampling procedure serves a uniquely different purpose. However, in all cases, a sample must be representative of the target population if inferences are to be made from the sample to the population.

The target population is the population to which researchers wish to generalize the results of a study. The target population must be a known and well-defined collection of individuals, events, or objects that is clearly specified. It behooves a researcher to know the attributes, features, or characteristics of the target population to assure that the sample is drawn from this group and to guarantee that only cases that fit into the specified population have opportunities to be selected for the sample. Cases that do not fit the constraints of the indicated group may not be representative of the target population. Care must be taken that such cases have a zero chance of being selected as part of the sample.

TABLE 1.2

Unique Features of Research Designs

Design	Type	Features
Experimental	Purpose:	Investigate cause–effect relationships
	Characteristics:	Random selection of subjects; random assignment of subjects to treatments or conditions; manipulation of independent variables; provides maximum control
	Advantages:	Maximum control; can claim cause–effect; replication and verification procedures
	Disadvantages:	Difficult to create pure experimental conditions in behavioral and social sciences research; experimenter expectations may bias interpretations
Quasi-experimental	Purpose:	Alternative to pure experimental design
	Characteristics:	May or may not have randomness; lacks control
	Advantages:	Useful for intact groups; can rule out rival hypotheses
	Disadvantages:	Risks internal and external validity; low statistical power
Descriptive	Purpose:	Investigate status quo
	Characteristics:	Random selection varies; no random assignment to treatments or conditions
	Advantages:	Useful for status quo studies; easy design for intact groups
	Disadvantages:	No treatment or control groups; no claim of causal relationships
Correlation	Purpose:	Investigate relationships between and among individual variables and sets of variables
	Characteristics:	Shows strength and direction (positive or negative) of linear and nonlinear relationships
	Advantages:	Easy to use and interpret; can serve as starting point for other analyses
	Disadvantages:	Does not show cause–effect relationships; no control over confounding variables
Causal comparative (ex post facto)	Purpose:	Investigate events or effects after they have occurred
	Characteristics:	Opposite of experimental design: causes sought after effects are observed; no random assignment to treatments or conditions
	Advantages:	Useful for retrospective studies; less costly than true experimental designs
	Disadvantages:	Lack of control; extraneous variables are possible

TABLE 1.2 (*Continued*)

Unique Features of Research Designs

Design	Type	Features
Longitudinal	Purpose:	Measure changes in a variable over time
	Characteristics:	Considered quasi-experimental or nonexperimental; randomization and level of control vary; measures made at different points in time
	Advantages:	Allows comparisons of one or more measures on one or more variables over time
	Disadvantages:	Internal and external validity affected by history, mortality, or attrition; usually costly and time consuming; low statistical power

For example, if one wishes to generalize to managers of large hotels, then the sample should be representative of managers of large hotels. Likewise, sound generalizations about librarians, school administrators, or counselors can be made only if a study sample represents the appropriate group.

A representative sample cannot assure the total absence of bias in a study, but representativeness is one major way that researchers can control for bias. Sample selection is a research design issue rather than a statistical one, even though the two are somewhat related. For example, different statistical procedures are appropriate for large sample sizes versus small sample sizes. Large and small sample sizes are defined by several aspects of a specific study such as the size and characteristics of the target population, research questions and hypotheses, types of data to be analyzed, and available resources.

Later in this section, sample size will be discussed. Keep in mind that the size of a sample cannot replace representativeness. Samples are selected in various ways to satisfy the purpose of the study and the need for representativeness. Samples may be classified as probability or nonprobability types. Probability sampling is based on the premise that each case in a population has an equal chance of being selected. Figure 1.1 shows the relationship of samples to available and target populations.

1.4.1 Probability Samples

Simple random samples are one of the most common types of probability samples used in research in behavioral and social sciences. Most researchers are familiar with random sampling. A stratified random sample includes cases selected randomly from each strata or subgroup of a population.

Samplings from intact groups such as classrooms are called cluster samples. Cluster sampling requires that all members of the cluster (all students in a classroom, all residents of an assisted living facility, all counselors in a school district) be selected. Intact groups are convenient samples for teachers, sociologists, psychologists, and others who may have access to special groups of individuals.

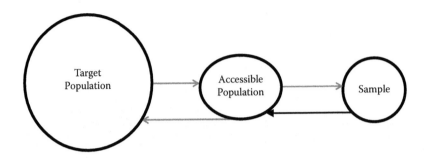

FIGURE 1.1
A random sample may be drawn from an accessible population.

Learning about the population constitutes a study in itself, not to mention the time and work involved in identifying all cases in a population. The list of all cases from which a sample can be selected is called a sampling frame. However, constructing a list of all cases in a given population such as individuals enrolled in higher education at any given time is next to impossible. Therefore, nonprobability sampling procedures can be used to select a sample.

1.4.2 Nonprobability Samples

Nonprobability sampling makes no guarantee that the selected cases are representative of the target population. Consequently, researchers must interpret the findings with caution. Quota samples and convenience samples are nonprobability samples. Quota sampling ensures that each segment or subset of a population is represented in the sample to the same extent that it exists in the population. When a quota is met, no more cases need to be selected. For example, if 20% of the children in a particular school are in the 10th grade, the researcher can set a quota to be sure that 20% of the sample includes children in the 10th grade. Such samples may or may not be representative of the target population when the target population includes all children in a school.

Convenience samples, as the name implies, are simply convenient cases that can be included in a survey. Convenience samples are accidental. For example, a researcher who interviews passersby in a mall or volunteers in a grocery store is using convenience sampling. This is a very weak method and is not recommended except in very difficult situations in which other sampling approaches are impossible.

1.4.3 Population and Sample Size

The most important characteristic of any sample is representativeness of the population from which it was drawn. Sample size affects the power of a test, and generally the larger the sample size, the more powerful the test. The relationship between sample size and statistical power is known as the *power efficiency* of a test. The power is the probability that a test will reject a null hypothesis when the null hypothesis is false.

Statistical power may be expressed as $1 - \beta$, where β (beta) is the probability of committing a Type II error or retaining a false null hypothesis compared to rejecting a true null (Type I error). As noted previously, the fewer and weaker the assumptions, the more general the conclusions that can be drawn. Hence, even when a statistical procedure requires few assumptions, power efficiency can be increased simply by increasing the sample size.

Unequal sample sizes can also affect the power of a test. A test may have higher power when the sample sizes are equal than when they are unequal. Formulas to calculate power efficiency are beyond the scope of this book. Suffice it to say that increasing the size of a sample generally increases the power of a statistical test.

Selecting an appropriate sample size is a major concern for both experienced and novice researchers. One way to choose an appropriate sample size is to decide on an acceptable margin of error and confidence level for hypothesized population proportions. Calculating a sample size using a desired margin of error and confidence interval is appropriate when a researcher is unsure of the size of the population. A formula that may be applied to calculate the sample size needed for a 10% margin of error and a 95% confidence level is

$$s = \left[\frac{1/10}{\sqrt{0.50(0.50)}} \right]^2 \times (1.96)^2$$

where s is the sample size, 1/10 is 1 divided by the desired margin of error (10% in this case), $\sqrt{(0.50)}\,(0.50)$ represents two hypothesized population proportions (0.50 was selected as the proportion that would yield the largest sample size), and 1.96 is the z score for the 95% confidence level.

Using the above formula, $s = 96.04$, which translates to 97 cases needed for the sample. Thus, one would be able to draw conclusions with 95% confidence and a 10% margin of error. When the number of cases in the total population is known, one can calculate a representative sample size based on the known population parameters. Using a similar formula, a representative sample size for any size population can be calculated:

$$s = \frac{X^2_{(1)} \times N \times (pq)}{d^2 (N - 1) + X^2 p(1 - p)}$$

where s is the sample size, $X^2_{(1)}$ is the chi-square value at one degree of freedom (3.84 from a chi-square table), N is the known number of cases in the population, p and q are the population proportions (generally set at 0.50 each to yield the largest sample size), and d^2 is the desired degree of accuracy expressed as a proportion.

Using the formula for a known population size of 354, we would need a sample size of 185:

$$\frac{3.84 \times 354 \times 0.25}{0.0025 \times 353 + 3.84 \times 0.50(0.50)} = \frac{340}{1.84} = 184.8$$

Researchers should always allow for a percentage of nonrespondents. Therefore, it is prudent to distribute more questionnaires or request more online participants than required by the formula.

It may not always be possible to draw a large sample. Sometimes a researcher may have only 15 or 20 cases available for a study. For example, small samples of 15 to 20 cases are common for small reading groups of elementary-level children within a school, secondary-level students in an accelerated mathematics course, students enrolled in specific kinds of special education programs, adult consumers of rehabilitation services, or identical twins within a specific age range. In such cases, a researcher may select a nonparametric statistical model. Nonparametric statistical models will be discussed in Chapter 2 and elsewhere in this book.

Sample size and power are important research design considerations. However, one cannot select a test with any degree of confidence without considering the validity and reliability of the measurement instruments to be used in a study.

1.5 Validity and Reliability of Measurement Instruments

Validity and reliability are key psychometric properties related to measurement instruments. The term *measurement* may be thought of as the assignment of numbers to an occurrence or observation such as an event, object, performance, knowledge, or attitude. When numbers are assigned in a meaningful way, researchers can use them to quantify and give meaning to their observations. Measurement instruments, tools, tests, and scales are all data collection devices. They are available in various formats and forms; examples are stop watches, weighing scales, tape measures, micrometers, thermometers, traditional paper and pencil tests, and electronic surveys.

Psychometrics is the field of study concerned with the development and testing of various kinds of tests and measurement scales. Measurement is not synonymous with evaluation or assessment. Measurement is concerned with the collection of objective data, whereas evaluation involves subjective interpretations or judgments based on data. Assessment is related to evaluation in the sense that assessment is performed to provide feedback to individuals as they progress through a course, task, or other activity, rather than to make a final appraisal of the work performed.

Instruments used in the social and behavioral sciences are expected to be valid and reliable. However, all data collection instruments are not necessarily tests of performance or knowledge in the traditional sense. For example, psychological tests of personality attributes, questionnaires, and attitude scales must also be examined for sound validity and reliability.

In most instances, validity and reliability are established in the development stages of tests. Procedures used to establish validity and reliability of commercially prepared tests are usually reported in a test manual or related literature. Even though psychometric elements have been established using sound and rigorous procedures, researchers should verify the validity and reliability of each instrument for their own unique research projects. In addition, researchers should confirm whether the types of validity and reliability established are appropriate for use in their studies. Only when valid and reliable instruments are used as measuring tools in research, can researchers expect unbiased and credible results.

1.5.1 Instrument Validity

Several different kinds of validity are common to measurement instruments used in behavioral and social sciences. Validity is usually defined as the extent to which an instrument measures what it purports to measure. However, the concept of instrument validity should be expanded beyond this simple definition to mean the extent to which valid decisions and predictions can be made as a result of data collected with a specific instrument. The following paragraphs present information related to the various kinds of validity and the different forms of reliability. An instrument cannot be valid unless it is also reliable; however, an instrument can be reliable without being valid. Reliability, which will be discussed later, means consistency; however, if the wrong variable is measured consistently, then an instrument cannot be considered valid.

Face validity is a qualitative decision related to the appearance, relevance, and representativeness of the items on a written instrument. Face validity is usually established by a 9- to 13-member panel of experts in research methods who review the instrument and all aspects of its appearance. The experts, sometimes called judges, assure that an instrument is indeed professional, user friendly, understandable, and easy to administer. Experts should check the overall format of an instrument for clarity of directions, appropriateness of response categories, usability, and meaningfulness of the items. Experts who provide valuable feedback help to avoid design problems during and after data collection. Instrument developers are expected to incorporate the corrections, changes, and additions suggested by the experts prior to administering the instrument. It is not uncommon for a panel to take a second look at an instrument after changes are made.

Content validity can be established by a panel of experts who verify the knowledge base of the items. The panel of experts assembled for face validity could include subject-matter experts to ensure that an instrument is ethical, professional, relevant, and representative of the content to be tested. Major concerns for content validity are completeness, correctness, and clarity of items.

Construct validity, on the other hand, is more elusive than face validity or content validity because constructs are often difficult to define. Unlike content validity that can survey a person's knowledge of a topic, construct validity addresses ambiguous concepts such as beauty, achievement motivation, imagination, and creativity. Construct validity is usually established through a series of statistical tests using correlation procedures and factor analysis.

Criterion-related validity refers to the extent that instrument scores or performance tasks are correlated highly with an existing test that is known to be valid. In the field of testing, a criterion is a standard or level of performance against which another performance is compared or correlated. Criterion-related validity can be established easily using item analysis, correlational techniques, and regression procedures.

1.5.2 Instrument Reliability

Reliability is the degree to which a measuring instrument is consistent over time on measures for similar populations. There are several different types of reliability, and each is a function of the purpose for which the results of the instrument are to be used.

Inter-rater reliability is the extent to which the ratings of two or more judges or raters are similar for an individual, object, or performance. Judges record their ratings and their ratings are correlated. A strong positive correlation denotes a high level of agreement between the judges.

Split-half reliability is the association between two halves of the same instrument to evaluate whether the items on both halves of a test measure the same factor. The halves of the test may be determined randomly, or all even numbered items can be compared with all odd numbered items to evaluate the strength of the correlation between the two sets of items. It is important that both halves of an instrument contain the same number of items at the same level of difficulty for similar content to ensure an accurate measure of the extent to which the two halves correlate.

Even the most careful procedures to establish reliability are not error free. Large differences in scores or values on one measure for individuals in two groups will generally produce a higher correlation coefficient than small differences in scores or values for individuals in two groups. This is true because individuals in groups with large differences will retain their rank within a group despite small changes in individual scores. For example, two groups of college freshmen could earn very different scores on the same sociology test. Perhaps one group was taught in a traditional classroom setting and the other was taught via distance education.

The same relationship is not true for individuals in a group with small differences among them, because small changes in each person's score or value can affect his or her rank in the group. For example, a small correlation would probably be observed for the heights of basketball players from two different teams. It is important to note that for groups with large observed differences and groups with small observed differences, only one measure is taken on all individuals in each of the two groups. Conversely, the correlation of two different measures for the same individual would yield a higher correlation coefficient when scores for the two measures are similar, such as measurements of self-concept and self-esteem. On the other hand, a low to moderate correlation may be observed between a student's score on a history test and his or her score on a calculus test. Table 1.3 illustrates the correlation of two related variables for an individual, and Table 1.4 illustrates the correlation between two unrelated measures for an individual.

TABLE 1.3

Scores and Correlations for Two Related Measures

		Measure 1	Measure 2
Measure 1	Pearson correlation	1	0.897[a]
	Sig. (two-tailed)		0.000
	N	10	10
Measure 2	Pearson correlation	0.897[a]	1
	Sig. (two-tailed)	0.000	
	N	10	10

[a] Correlation is significant at 0.01 level (two-tailed).

TABLE 1.4

Scores and Correlations for Two Unrelated Measures

		Test 1	Test 2
Test 1	Pearson correlation	1	0.050
	Sig. (two-tailed)		0.890
	N	10	10
Test 2	Pearson correlation	0.050	1
	Sig. (two-tailed)	0.890	
	N	10	10

1.6 Steps of the Research Process

Research should be conducted in a systematic manner so that the completion of each step provides a basis for the next. Although researchers may move back and forth among the early stages of research, for example to add new information or to refine terminology, the basic steps should be completed methodically. This section presents the steps in the research process that address the basic components of research. These steps provide a format that is useful in the formulation of a research proposal to investigate a topic of interest. Furthermore, the proposal format can be used as an outline for a research report because it provides a reader with essential information needed to evaluate the process and to subsequently interpret the results of the investigation. The following steps illustrate the scientific method for conducting research in the behavioral and social sciences.

1.6.1 Introduction

The introduction provides background information and establishes a framework for a study. The introduction should include a theoretical and/or empirical basis for the topic and the study. The focus of the study should be established clearly to give readers insight into the direction of the study. Contrary to some views, there is no shortage of researchable problems in the behavioral and social sciences. The field is rich with ideas that can be explored; however, novice researchers generally do not have the practical experience to select suitable problems. Therefore, they must rely on the related literature, completed research, professional contacts, or opportunities to replicate existing studies.

1.6.2 Statement of Research Problem

An initial step in the research process is identifying and stating a problem to be investigated. Practitioners encounter myriad problem choices during their daily work. For the student or novice researcher, however, selecting a problem is often overwhelming. The best place to start problem identification is with professional contacts. College faculty members, job supervisors, and others can suggest problems for study. A researcher with limited professional experience can review the related literature and existing research to find topics of interest. The problem for study should be stated succinctly and clearly early in the introduction.

The best research problem statements are expressed in one sentence. One or more lead-in sentences can be used before the actual problem statement; however, the problem statement should express clearly the intent of the research and the relationships to be explored. Rather than have readers guess the focus of a study, some researchers will state that the problem is a lack of information about a given topic that they wish to investigate. Writing a good research problem statement is difficult because the statement must show that the researcher proposes to investigate relationships, effects, or differences about the research topic. For example, a problem statement for a research study designed to explore teacher satisfaction could be stated as: The focus of this research is the lack of information related to teachers' job satisfaction in elementary-level schools. After the problem is identified and stated clearly, the researcher should gather information (research and relevant literature) related to the problem.

1.6.3 Review of Related Literature

Regardless of the source of a research problem, a complete review of the literature is necessary to develop background information and a theoretical basis for the study. The purpose of the study and the need for the investigation can be supported by existing literature. Possible research questions can be derived from a review of literature. A thorough review will show a researcher where the gaps are in the knowledge of a specific topic. A literature review should follow a systematic and logical progression of information. For example, a review can be organized chronologically by dates or time periods or topically by subject areas.

1.6.4 Statement of Limitations

Limitations help define broad terms such as quality of life, job satisfaction, and beauty within the context of a specific study. Limitations related to the conduct of a study, such as sample choice, sampling procedures, research instruments, methods of data collection, return rate of survey forms, and analyses should be addressed. Researchers should acknowledge that certain limitations may compromise the validity of their findings and limit the extent to which results can be generalized to a target population. Some limitations can be remedied during the research design phase but that does not apply to assumptions.

1.6.5 Statement of Assumptions

Statements or ideas about a study that a researcher accepts as truth are called assumptions. For example, a basic assumption of all survey research is that participants will respond honestly to the items. Obviously, researchers cannot claim that this is a true assumption; however, the fact that it exists should be stated. Generally, researchers can use reasoning, empirical evidence, authority, or previous studies to support their assumptions. Caution should be exercised when stating assumptions. Checking assumptions against logic, evidence, authority, or existing literature is a good place to start.

1.6.6 Formulation of Research Questions

After a thorough review of the literature is completed, specific research questions can be formulated. One or more research objectives or questions may grow out of the problem statement. Research questions should be relevant and meaningful to the topic so that their responses contribute to the solution of the problem. In behavioral and social sciences research, it is wise to avoid "yes or no" research questions. Questions that elicit "yes or no" responses do not provide sufficient information for researchers to formulate informed conclusions about findings. When researchers report the results for "yes or no" questions, results are tabulated and further discussion goes beyond the research question.

While this may seem a minor point, it is a practice to be taken seriously. Researchers should report their findings in such a way that a discussion of findings is consistent with the way in which the question was asked. For example, in the teacher job satisfaction example cited above, appropriate research questions are

What are the demographic characteristics of teachers related to their age, level of education, and years of teaching experience?

To what extent is teacher job satisfaction related to salary, administrator support, and parental involvement?

Trivial questions about teachers' hair colors or lunch choices should be avoided. Responses to research questions should provide information that adds to the solution of the problem. An exhaustive review of the literature demonstrates that a researcher has a sound understanding of the problem and the gaps in existing information. Questions arise when the literature review suggests that more information is needed about a topic. Some research questions must be tested statistically before the answers can be known. At this point, hypotheses should be developed.

1.6.7 Development of Hypotheses

After research questions are specified, testable hypotheses can be formulated. A hypothesis is a statement of a tentative theory or supposition adopted for the purpose of explaining certain known conditions. Hypotheses guide the research process and are usually classified as directional or nondirectional.

A directional hypothesis states a researcher's prediction of the effect, relationship, or difference among variables. For example, a directional hypothesis could state that motivation achievement scores on the ABC test of comprehension are higher for children whose parents are involved in their school work than children whose parents are not involved. A nondirectional hypothesis makes no such prediction. Nondirectional hypotheses are stated in the null form. A null hypothesis (H_0) can be tested statistically for results that fall within a preset probability level. The following is an example.

H_0: no difference exists in achievement motivation scores on the ABC test of comprehension between children whose parents are involved in their children's school work and children whose parents are not involved in their school work.

Not all research designs require hypotheses. For example, hypotheses have no function in descriptive studies. Hypotheses should be stated in clear, concise, and understandable language for the specific topic to be studied. A controversy among researchers about the best way to write hypotheses has been ongoing for several decades. Some researchers assert that the directional hypothesis that forces them to make a prediction can be written more clearly than null hypotheses and directional hypotheses indicate a deeper understanding of a problem. Other researchers who support use of the null hypothesis contend that directional hypotheses invite researcher biases that may carry over to the conduct of a study. In classical research, use of the null form is still encouraged to assure researcher neutrality and fairness in hypothesis testing.

1.6.8 Populations and Samples

A well-defined population should be identified as a target for a study. A target population is a group of individuals, events, or objects to which researchers wish to generalize results. All individuals within a target population may not have the same opportunities for selection, especially when national groups (all students in the eighth grade, all former college administrators, or all retired athletes) comprise a target population. In such situations, researchers may select a sample from a large available population (students in the eighth grade, college administrators, or retired athletes) in a specific geographic location. Section 1.4 discussed populations and sampling techniques for a representative sample. After a sample is selected, one may move to selecting or developing research instruments and data collection.

1.6.9 Research Instruments

A researcher should preplan and document all instruments and procedures used during the data collection process. Researcher-developed or commercially prepared instruments may be used. All related information such as name of the instrument, author or publisher and date, purpose, reliability and validity, procedures for establishing reliability and validity, and any other pertinent information such as procedures used to develop the instrument should be reported.

1.6.10 Data Collection

Data may be collected via several methods such as open-ended or structured interviews and mail-out or online survey forms. Enough time should be allotted for respondents to participate, and follow-up requests should be utilized for nonrespondents. Usually follow-up requests are made two weeks after an initial request for participation. It is not unusual to conduct two follow-up requests.

After the data are collected via mail and online surveys, they should be screened for complete and useable responses. Inventories, questionnaires, and opinionnaires should be examined and forms with many missing values or forms that appear to have set responses should not be included. Depending upon the statistical procedures to be used, the data may need statistical screening as well. It is a good idea for researchers to number each form so that it can be reviewed easily if a mistake is made when data are entered into a spreadsheet for analysis.

1.6.11 Analyses

After useable data are entered into a spreadsheet, statistical analyses can be conducted. Specific analysis procedures should have been decided when research questions were formulated. Research questions and hypotheses drive an analysis. For example, a hypothesis stating that no relationship exists between variables needs a correlational analysis. A hypothesis stating that no differences exist between groups requires statistical procedures to detect differences.

1.6.12 Interpretation of Results

Interpreting results and drawing conclusions should be based solely on the evidence. Moving beyond the evidence to make conclusions will prompt enlightened research consumers to question not only the conclusions but also the conduct of the study in general. However, researchers are not tied exclusively to the statistical results in their discussions of implications of the findings. It is acceptable and expected that researchers include a discussion of the meaning and implications of the results to stimulate additional research questions for further study.

1.6.13 Statement of Conclusions

Conclusions should be based on statistical findings. Possible applications to professional practices can be included. The conclusions should hold some implications for the improvement of situations related to the problem investigated.

1.6.14 Preparation of Final Report

The last step is preparing a final report that should include a brief abstract and a summary of all important components of the study, along with the findings, conclusions, and a brief explanation of the results. Appropriate charts and graphs are always useful in a final report. Figure 1.2 lists the steps for using a scientific method to conduct research.

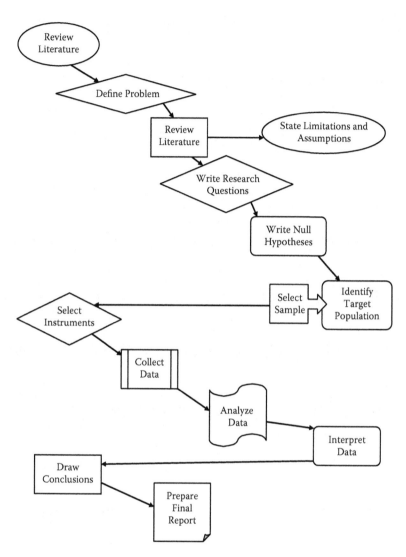

FIGURE 1.2
Steps of research process.

Chapter Summary

Teachers, managers, administrators, and other professional personnel are faced with questions to answer and problems to solve. For example, school board members, parents, and business and industry personnel look to research to understand certain phenomena such as job satisfaction, school violence, achievement motivation, and workplace values. Questions and problems often require in-depth study before effective solutions can be determined.

Why do some individuals achieve in the workplace? Why do some learners drop out of school? What is the most effective method for teaching mathematics? Do early readers become overachievers? These are examples of the kinds of problems that lead to fruitful research investigations. To answer these questions, the appropriate practice is to design and conduct an investigation that will provide acceptable solutions. Therefore, the initial steps in conducting a research study are critical to sound outcomes.

Basic research is based on empirical evidence that supports a current theory or suggests a new one. Applied research is more practice-oriented and designed to solve immediate problems. Experimental designs use randomization, control, and manipulation to rule out plausible explanations (other than the cause of interest) for the effects of treatments or differences in outcomes under different conditions. A true experimental design allows researchers to attribute changes in behavior to a condition or treatment. On the other hand, quasi-experimental designs do not impose the same rigorous requirements. Quasi-experimental designs are popular for studies of intact groups in situations where strict controls are impossible.

Nonexperimental designs include a category of studies that lack controls such as random assignments of subjects to treatments. Such designs are important to describe the status quo or show relationships as in descriptive and correlational studies, respectively. Retrospective studies that search for causes of effects are classified as causal comparative designs. These designs do not meet the strict criteria of experimental or quasi-experimental designs. Longitudinal designs are useful for following changes in measures over time. Problems associated with prolonged time periods and extraneous variables are typical for longitudinal studies.

Regardless of the sampling method, the top criterion for sample selection is representativeness of the target population. Probability samples help assure that each case in a population has an equal chance of selection; nonprobability samples make no such claims.

Validity and reliability of measuring instruments must be verified for the specific study for which they are used. Valid and reliable instruments assure the integrity and credibility of a study and also minimize researcher bias.

In practice, the development and progress of a research study does not occur in isolated steps. Procedures that follow lock-step methods in the early stages of research are far too restrictive and mechanical to be realistic. The first steps involve reviewing literature and identifying a problem to be investigated. The research problem statement should note that evidence covering the topic is inadequate. Thus, the purpose of research is to provide evidence that will contribute to the existing body of knowledge and lead to improved practice and decision making. Research questions should be written in the early stages of the research project.

Tasks for the remaining steps should be completed in the specified sequence. These steps include formulating hypotheses, selecting a sample, selecting or designing research instruments, collecting and analyzing data, interpreting the data, drawing conclusions, discussing the results, and preparing a final report.

A study that lacks meaning by failing to contribute to an existing theoretical base or application to real-life problems cannot be justified. Quantitative studies provide objective evidence related to a problem rather than a researcher's philosophical views. Studies that lack objectivity are inefficient and unwise undertakings in an economy where time and cost effectiveness are paramount.

Evidence-based practices require that research be carried out in the most unbiased and objective manner possible. Sound research designs help to assure credibility of results and believability of conclusions and recommendations. Studies that follow the scientific method should be replicable by other researchers. Replication is necessary for results to be verified.

This chapter provided background information on the purpose of a research plan and explained major research designs, along with their advantages and disadvantages. Steps in the research process were outlined. Conducting a statistical procedure without a context for a study is meaningless. Data analysis was mentioned only briefly in this chapter. Succeeding chapters will present specific categories of nonparametric statistical methods for data analyses. The design of a study, type of data, sample selection, research questions, and hypotheses are critical to the type of statistical analyses needed. Various statistical analysis procedures are available to researchers. Statistical procedures may be classified as parametric or nonparametric; each category has its own specific uses and techniques.

Chapter 2 introduces the development and distinctiveness of nonparametric statistics. Chapters 3 through 8 discuss a variety of nonparametric tests commonly used in social and behavioral sciences research.

Student Exercises

Complete the following exercises.

1.1 Locate three refereed research articles that report results of experimental studies. Select articles published in a journal for your field of study. Critique each article using the following guidelines.

a. Give the title of the article, author(s), name of journal, volume, issue number, and page numbers.

b. What research problem was addressed?

c. What were the research questions or objectives?

    d. Who or what was the target population?

    e. What procedures were used to select the sample?

    f. What were the independent variables? What were their levels?

    g. What was the dependent variable?

    h. What was or were the treatment(s)?

    i. In what way were the independent variables manipulated?

    j. What was or were the hypothesis(es) tested?

    k. What were the conclusions of the research?

    l. Write a statement about the overall quality of the research design and the conclusions drawn.

    m. What indicators demonstrated that the study was actually experimental?

1.2 Locate two refereed research articles that report results of a quasi-experimental study. Select articles published in a journal for your field of study. Critique each article using the following guidelines.

    a. Give the title of the article, author(s), name of journal, volume, issue number, and page numbers.

    b. What research problem was addressed?

    c. What were the research questions or objectives?

    d. Who or what was the target population?

    e. What procedures were used to select the sample?

    f. What were the independent variables? What were their levels?

    g. What was the dependent variable?

    h. At what point in the study was a treatment, if any, applied? What was the treatment?

    i. What was/were the hypothesis(es) tested?

    j. What were the conclusions of the research?

    k. Write a statement about the overall quality of the research design and the conclusions drawn.

    l. What indicators demonstrated that the study was quasi-experimental?

1.3 Locate two refereed research articles that report results of a non-experimental study. Select articles published in a journal for your field of study. Critique each of the articles using the following guidelines.

    a. Give the title of the article, author(s), name of journal, volume, issue number, and page numbers.

    b. What research problem was addressed?

    c. What were the research questions or objectives?

    d. Who or what was the population?

    e. What procedures were used to select the sample?

    f. What were the conclusions of the research?

    g. Write a statement about the overall quality of the research design and the conclusions drawn.

    h. What indicators demonstrated that the study was nonexperimental?

1.4 Review the sampling procedures discussed in this chapter. Which are the most appropriate to use in your field of study? Why?

1.5 Develop a research plan on a topic that is appropriate for your field of study. Provide responses for each of the following.

    a. Review published research and related literature on a topic of interest in your field of study and prepare a ten-page summary of your review.

    b. Write a research problem statement based on your review of the research and literature.

    c. If you were to conduct a study on your selected topic, which research design would you use? Why?

    d. Write research questions based on your problem statement in item b.

    e. If you need to test hypotheses (based on your design and questions), write the hypotheses.

    f. Which procedures would you use to select a sample?

    g. If your design allows, which variables would you manipulate?

References

Ary, D., Jacobs, L. C., Razavieh, A., and Sorensen, C. K. (2009). *Introduction to Research in Education*, 8th ed. Belmont, CA: Wadsworth.

Babbie, E., Halley, F., Wagner III, W., and Zaino, J. (2011). *Adventures in Social Research: Data Analysis Using IBM SPSS Statistics*, 7th ed. Thousand Oaks, CA: Sage.

Black, T. R. (2009). *Doing Quantitative Research in the Social Sciences: An Integrated Approach to Research Design, Measurement, and Statistics*. Thousand Oaks, CA: Sage.

Daniel, J. (2012). *Sampling Essentials: Practical Guidelines for Making Sampling Choices*. Los Angeles: Sage.

Ramsey, F. L. and Schafer, D. W. (2012). *The Statistical Sleuth: A Course in Methods of Data Analysis*, 3rd ed. Pacific Grove, CA: Duxbury Press.

Tabachnick, B. G. and Fidell, L. (2007). *Experimental Design Using ANOVA*. Belmont, CA: Thomson, Brooks/Cole.

Tanner, D. (2012). *Using Statistics to Make Educational Decisions*. Thousand Oaks, CA: Sage.

2

Introduction to Nonparametric Statistics

Chapter 1 presented an introduction to research. The basic principles of research and planning for research were discussed, along with explanations of various types of research designs. Information about sampling procedures, different kinds of probability and nonprobability samples, and calculating sample sizes were included. Considerations for the validity and reliability of research instruments were described. Finally, steps in the research process were delineated with explanations of each step.

A solid research design is necessary to provide a basis for trustworthy results of an analysis. Otherwise, the statistics provide no context within which to make interpretations and draw conclusions. Without a context for interpretation, statistical results are meaningless.

This chapter extends the discussion of research planning to information related to statistical procedures. It begins with an introduction to data analysis, followed by an overview of nonparametric and parametric statistics. Next, a discussion of specific requirements of parametric and nonparametric statistics is provided. The discussion of parametric statistics is included in this chapter because these statistics are used often by researchers. An overview of parametric statistics is intended as a point of reference to help students place nonparametric statistics in perspective. In addition, measurement scales are discussed and illustrated. The chapter concludes with a summary, followed by exercises to reinforce learning and encourage students to apply the concepts and principles covered in the chapter.

2.1 Data Analysis

Data analysis, as the name implies, is the process of analyzing data. The analysis may be simple or complex, depending upon the research questions and the hypotheses to be tested. Data analysis begins officially after the data are collected, screened, and entered into a computer program for analysis. Generally, statistical procedures used for analysis may be classified as descriptive or inferential.

Descriptive statistics are useful for organizing and summarizing data and provide a "picture" of the data. For example, frequency of occurrence of each score in an array of scores, average score among a group of test takers, the spread of scores about an average or mean score, and a range of scores may be ascertained by descriptive statistics. In addition, correlation procedures are used to describe the association between and among sets of variables.

Inferential statistics: Simply reporting descriptive statistics about a group or groups may be insufficient. Sometimes, it is important and necessary to examine differences between and among the mean scores of groups so that inferences can be made from a smaller group (sample) to the population of interest. In such cases, researchers use inferential statistical procedures. Procedures such as the t-tools, analysis of variance, regression, chi-square, and the Kruskal-Wallis test are some of the more common inferential statistics. Many different kinds of inferential statistical procedures are useful to explain data that result from research processes.

2.2 Overview of Nonparametric and Parametric Statistics

Statistical procedures may be grouped into two major classifications: parametric and nonparametric. Specific statistical tests are available within each category to perform various analyses; however, data for each category should meet unique requirements prior to analysis.

Parametric statistics require assumptions to be more specific and more stringent than the assumptions for nonparametric statistics. The generality of conclusions drawn from statistical results are tied to the strength of the assumptions that the data satisfy. In other words, the more rigorous the assumptions, the more trustworthy the conclusions. Conversely, fewer or weaker assumptions indicate weaker or more general conclusions. For this reason, statisticians prefer to use parametric tests whenever the data meet the assumptions.

Unlike parametric statistics, nonparametric statistics make no assumptions about the properties of the distribution from which the data are drawn, except that the distribution is continuous. The properties of a distribution are defined by its shape (normality), midpoint (mean), and spread of scores (variance and standard deviation). These measures represent the parameters of a population. Before conducting statistical tests, researchers generally check the assumptions of the data to verify whether they are appropriate for a parametric test.

Parametric tests are more powerful than nonparametric tests in that they have a better chance of rejecting the null hypothesis when the null hypothesis is false. For example, if one were to test the null hypothesis for mathematical achievement of students who are taught beginning algebra using three different instructional methods, one would want to use the most powerful statistical test available.

Since parametric statistical tests are more powerful than nonparametric tests, the immediate choice for researchers is to use a one-way analysis of variance procedure to test for possible differences among the groups. If differences are revealed, follow-up procedures using pairwise comparisons may be conducted. Analyses of variance procedures are parametric; therefore, they require that certain assumptions be met before researchers can have confidence in the results.

Statistical tests are available to test some assumptions; others should be incorporated into the research design. Different kinds of assumptions should be met, depending upon the level of complexity of the research design and the statistical procedures. Basic assumptions are normality, homogeneity of variance, randomness, and independence. If either the normality or equal variance assumption is violated, whether the assumptions of randomness and independence are met is a moot point. Both parametric and nonparametric procedures are used for hypothesis testing. Parametric statistics are used often in social and behavioral sciences research; thus, an overview of parametric statistics may be helpful as a starting point before moving directly into a discussion of nonparametric statistics. (See also the comparison in Table 2.1.)

2.3 Overview of Parametric Statistics

Parametric statistics constitute a group of tests appropriate for analyzing interval and ratio data. Parametric statistics are often referred to as inferential because they assess parameters of a population based on statistics from

TABLE 2.1

Comparison of Nonparametric and Parametric Statistics

Nonparametric Statistics	Parametric Statistics
Continuous distribution	Assumptions of normality and equal variances
Uses median as location parameter	Uses mean, variance, and standard deviation as location parameters
Random sample	Random sample
Independence of responses	Independence of responses
Uses nominal, ordinal, interval, and sometimes ratio data	Uses interval and ratio data
Large and small data sets	Large data sets (minimum of 30 or more cases)
Weaker statistical power than parametric statistics	More powerful than nonparametric tests for rejecting null hypothesis

a sample. Parametric statistics are not distribution free; that is, sampling distributions are based on the population distribution from which the samples were drawn.

Parametric procedures are based on several important assumptions. One assumption is that the population data are normally distributed, and sample data taken from that specific population are also normally distributed. A normal distribution has data points that fit symmetrically on a bell-shaped curve. Generally, parametric procedures are robust to departures from normality. However, depending on the magnitude of a departure, incorrect results can lead to incorrect conclusions. In such cases, transforming the data using higher-order statistics such as cubes or square roots may create a more normal appearing distribution.

After transformations are made, an analysis can be conducted using the higher-order statistics. Normality and the inferential nature of parametric statistics allow researchers to conclude within a certain probability the chances that a score falls within or outside a certain range of scores. Drawing a random sample from a normally distributed population helps to meet the normality assumption. In addition, the mean and standard deviation of a sample are important in making inferences.

The mean is the average score for a group, and the standard deviation indicates the spread of the scores around a mean value. The standard deviation shows the average distance of individual scores from a group mean. These parameters allow researchers to predict the probability of a certain score occurring by using the z table for a normal distribution. The more normally distributed the data, the more accurate the results of the analysis. For example, measures for two independent groups may reveal similar median values, but the mean values may be vastly different. In such cases, the data may not be normally distributed and a parametric procedure could yield a statistically significant result, whereas nonparametric results may not be statistically significant.

Another basic assumption of parametric statistics is homogeneity of variances. Homogeneity of variance requires that the data from different comparison groups have a similar spread of scores around the mean. This means that the variances for the comparison groups should be similar in all important aspects to the analysis. Some parametric procedures are robust to unequal variances. When the assumption of equal variances is violated, most statistical software reports the test for unequal variances along with the test for equal variances.

Parametric statistics are based on interval or ratio data; therefore, data analyzed using parametric procedures should be from a true interval (continuous) or ratio scale. A true interval is one in which distances or degrees between the points on the scale represent equal intervals. Since parametric statistics use higher-order scales of measurement (interval and ratio), data can be changed to a lower-order (nominal and ordinal) scale.

Such conversions may be useful when an independent variable such as income is measured on an interval scale and a researcher wishes to place respondents into high, medium, and low income groups. Parametric statistics are more powerful than nonparametric statistics; consequently, parametric statistics may yield accurate results with normally distributed small sample sizes for most procedures.

Finally, two remaining assumptions for parametric statistics are independence of observed scores and randomness. Independence means that a score that a participant receives on any item in no way influences the score of any other participant on that item. The assumption of randomness means that the subjects or cases for a study were selected randomly, and the selection of one subject or case for a study is totally independent of the selection of any other subject or case for the same study. The assumptions of independence and randomness are not tested statistically. They should be met as part of the research design. Regardless of the research design or statistical procedures applied, if the assumptions of independence and randomness are violated, the results are biased at best and also flawed and untrustworthy.

2.4 Overview of Nonparametric Statistics

The development of nonparametric statistics can be traced to the early 1700s when Arbuthnott, a Scottish-born physician and author, used binomial probability calculations. New nonparametric procedures, along with modifications, refinements, and extensions of earlier procedures have been developed over the last 60 years. Early developments and more recent ones are discussed at appropriate points in this book when specific procedures are presented.

Nonparametric statistics involve a group of procedures known as distribution-free statistics. Nonparametric procedures should meet certain assumptions, as do parametric procedures, although the assumptions are different. The data for nonparametric analyses do not need to be normally distributed as required for parametric or classical statistical procedures. Recall that for parametric statistics, a distribution is assumed to be normal, with a mean of zero and equal variances.

Specifically, the distribution of a sample selected for a nonparametric analysis does not depend on the distribution of the population from which the sample was taken. In other words, interpretations of nonparametric data do not depend on the fit of the data to any parameterized distribution. While the assumptions of normality and equal group variances are imposed on parametric statistical procedures, the major requirement for nonparametric data is that the data are independent and identically distributed (i.i.d.), that is, data are independent and come from an identical continuous distribution. Data for nonparametric analyses may be based on ordinal, nominal, interval, or ratio scales.

Nonparametric procedures are appropriate alternatives for small data sets or when interval or ratio data do not meet the assumptions necessary for parametric statistics. At the same time, small samples make nonparametric statistics less powerful than parametric statistics, particularly in situations where data meet the normality assumption. Nonparametric procedures are often thought to be simpler than their parametric counterparts.

Generally, nonparametric statistics are easier to calculate than parametric statistics. Even so, results from nonparametric statistics can be nearly as accurate as their parametric counterparts, particularly when a distribution is normal. For example, a sample of 75 or more cases can be analyzed successfully with parametric and nonparametric methods, so long as the data are interval. However, if the data are ranked, parametric procedures are inappropriate. When a distribution is nonnormal, nonparametric statistics are considered to be more efficient, as nonparametric procedures are insensitive to outliers.

Exact p-values for nonparametric statistics can be obtained easily by using computer software for large or small sample sizes. Prior to the development of statistical software to perform the calculations for exact p-values, desk calculators were used. However, as one might imagine, hand calculations involve lengthy processes and are prone to error. Large sample sizes yielding approximate p-values are preferred over hand-calculated exact p-values.

2.5 Importance of Nonparametric Methods

The use of nonparametric statistics for small sample sizes is an advantage for researchers when it is difficult or impossible to gather sample sizes of 30 or more participants. Some clinical studies collect interval data; however, large data sets are not always obtainable and thus nonparametric procedures are used. For example, studies of biostatistics may use small data sets. In addition, data from many disciplines and areas of study are often dichotomous on both the dependent and independent variables.

As an example, an educational researcher may want to investigate the extent to which differences in performance as pass or fail outcomes (dichotomous dependent variables) exist for boys and girls (independent variable) who received or did not receive instruction on anger management (independent variable). In this case, all variables are dichotomous.

Another important application of nonparametric statistics is for analysis of ranked or ordered data. Ranked or ordinal data are collected frequently in disciplines such as kinesiology, rehabilitation, business, and medicine. For example, participants may be asked to rank their responses from one time, condition, experience, or preference to another.

Ranks are often used in contests in which individuals compete for first, second, and third place. A person who wins a swim meet may be ranked as number 1 for first place; the person who finishes the meet second is ranked as number 2 for second place, and the person who finishes the meet third is ranked number 3 for third place, and so on. The rankings of first, second, and third place indicate only the order in which the swimmers finished the contest. The time lapses between each swimmer finishing the race most likely differ, that is, the time lapse between the first and second place winner is most likely not the same as the time lapse between any other two swimmers who finish the meet.

Sometimes it may be desirable for cases to be ranked even though the data are continuous in nature. For purposes of measurement, interval data used in nonparametric procedures are calculated as ranked data. However, ordinal data are not treated as interval data for parametric procedures.

Many different types of nonparametric methods are used in hypothesis testing. One group is categorized as location tests that make statistical inferences by testing the null hypothesis on the value of a measure of central tendency (median) and estimates of confidence intervals. Common hypotheses tested in social and behavioral sciences research are related to the central tendency of the data, patterns of scores for single samples, independence of two or more samples, and the magnitude of difference between dependent samples. Also, tests of correlations and associations are performed using nonparametric statistics. Location tests are appropriate for large and small sample sizes.

Thus far, an introduction to parametric and nonparametric procedures and the requirements for each category of procedures were discussed. A large part of the discussion focused on the type of data appropriate for analysis using each category of procedures. Regardless of the statistical methods used, measurement instruments are needed to collect data for analysis; they are discussed in the next section.

2.6 Measurement Instruments

Researchers in the social and behavioral sciences do not have the luxury of the precise measurement instruments that exist in the physical and natural sciences. For example, job satisfaction, aptitude, attitude, achievement, learning, and motivation parameters all present measurement challenges. First, the variables are difficult at best to define. Individuals have different perspectives of constructs related to human behavior. Second, capturing the true essence of the meaning of constructs with a single or even several measurement instruments is difficult and often open to question.

Social and behavioral measures usually reference human behavior; and human behavior is complex and multifaceted. Since measurement in the behavioral and social sciences is difficult, is frequently inexact, and may cause harm if inaccurate conclusions are drawn, one might question why researchers bother with such measures. The point is that measuring instruments have become more highly developed and finely tuned over time. Failing to conduct research leaves a void in progress toward finding solutions for many social, political, and economic problems of the 21st century. Intelligent and informed decision making is based on information; consequently, it is appropriate and necessary that researchers continue to develop and refine instruments to specify and measure complex behaviors.

Many of the variables used for behavioral and social sciences research pertain to education; however, the concepts of measurement apply equally well in other environments such as biological research, industry, business, government, and geographic locations. Measurement scales are important in data analysis because statistical models and resulting data interpretations are dictated by the types of measurement scales used.

S. S. Stevens, a Harvard psychologist, introduced the notion of a hierarchy of measurement scales in 1946. Since then, scales have been the centers of controversies among statisticians and practitioners alike. Nonetheless, measurement scales are important for statistical analysis because the type of measurement scale used to collect the data influences analyses and resulting interpretations of the data. For example, data collected on an interval scale are amenable to many statistical procedures; whereas analyses are limited for data collected on a nominal scale.

In the remainder of this section, measurement scales proposed by Stevens are discussed. The discussion proceeds from the most basic or least sophisticated scales to the most complex or sophisticated.

2.6.1 Nominal Scales

Nominal scales record data at the lowest level of measurement for statistical purposes. They are naming scales and considered the least complex of all measurement instruments. Data collected on nominal scales are expressed as categories such as religious affiliation, year in school, favorite college course, etc.

Nominal scales may collect dichotomous data such that participants are required to select one and only one response from two possible options (yes or no; pass or fail; happy or not happy). Some nominal scales involve more than two categories such as school class (freshman, sophomore, junior, senior); marital status (single, married, divorced, widowed); or favorite pastime (watching television, reading, playing sports, visiting friends).

Statistical procedures that can be applied to nominal data are limited because these data have no quantitative value. For example, favorite school subjects may appear in the following order on a nominal scale: history, literature, statistics, and philosophy. The order in which the courses are listed on the scale is unimportant. Likewise, the numbers used to code the categories are not quantitatively important. Participants would not need to see the coding because the coding is used only for data entry into a computer spreadsheet. If literature is coded as a 2 for example and statistics is coded as a 1 for analysis, the 2 does not mean that literature has twice the value of statistics. Frequencies and percents are usually calculated to show the frequency of participants who selected each subject and the percent who selected a subject from the total group.

2.6.1.1 Scale characteristics

Nominal variables are coded with numbers so that they may be treated statistically. For example, the "preferred pet" variable may be coded by assigning 1 to cat, 2 to dog, and 3 to horse. The categories are mutually exclusive. Numbers are used to create *dummy* variables. The numbers themselves are arbitrary and meaningless. The numbers identifying each of the categories can be interchanged without changing the properties of the scale. A person who prefers a dog to a cat does not have twice as much preferred pet level as a person who prefers a cat. It is necessary to assign numbers to the categories only to allow the variables to be identified in a computer spreadsheet and then organized and summarized.

Statistical inferences cannot be drawn from nominal data. Nominal data are not amenable to parametric methods. Typically, the only kind of analysis that can be conducted is counting the number of cases and reporting the frequency and percent of occurrence of a given variable in each category. Since nominal scales express no values or relationships among the variables, no mean can be computed. The only meaningful measure besides the frequency and percent would be the mode, which indicates the most frequently occurring category. Figure 2.1 displays a nominal scale.

Controversy continues whether nominal variables can be measured on a scale at all, since nominal variables do not exhibit scale qualities such as magnitude or equal distances. However, nonparametric procedures are available for further analysis of nominal data. These procedures will be discussed in later chapters.

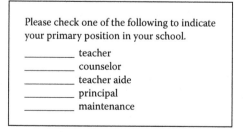

FIGURE 2.1
Example of nominal scale.

2.6.2 Ordinal Scales

An ordinal scale is useful for ranking qualitative variables such as the order in which students complete assignments, athletes finish a competition, adults list their preferences for certain foods, or order in which clients rank the efficiency of services they received. Scores on standardized tests are often presented as ordinal data. For example, percentile rank in a high school class may be used by college admission boards. Military ranks and socio-economic status are other examples of ordinal scales. Survey forms often include sections that ask respondents to rank some aspect of the variable being measured. When a list of objects can be ranked from highest to lowest or lowest to highest, an ordinal scale is appropriate.

Unlike numbers on a nominal scale, the numbers on an ordinal scale have meaning. The numbers assigned to the objects to be ranked order the objects in the same way that the attribute, preference, or scores are ordered. Careful attention should be given to the order of numbers on an ordinal scale to facilitate the ease and accuracy of interpretation of the data. For example, 1 does not always mean first; 2 does not always mean second, and so on. A 10-point scale may be presented when asking respondents to rank their perceptions of a service or product. A rank of 10 may be interpreted as the most positive perception and a rank of 1 may be interpreted as the least positive perception. Conversely, if people are asked to rank their preferences for various foods on a 10-point scale, a rank of 1 may be interpreted as first choice, 2 may be interpreted as second choice, and so on. Information gathered from the ranking influences the way in which the scale is constructed.

2.6.2.1 Scale characteristics

The only measurement property that ordinal scales exhibit is order of objects. As long as the order of the objects does not change, any ranking system is acceptable. The ranks alone do not give information about the amount of the item ranked. The ranked objects may fall on a continuum that specifies the location or category of an object at a specific point on the continuum. For example, a sprint runner who finishes a race in third place is behind the runners in second and first places, but this does not mean that the third-place runner took three times as long to complete the race as the runner who finished in first place or that the runner in second place took twice as long to finish the race as the runner in first place. The distance between the numbers does not have constant meaning. Only order of completion is known.

Another example is the satisfactory–unsatisfactory scale for scoring performance on a test. The underlying continuum that specifies whether a score is in the satisfactory or unsatisfactory category may be from 0 to 100, with a score of 70 or less on the scale indicating unsatisfactory performance and a score greater than 70 indicating the satisfactory performance category.

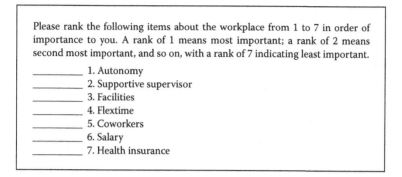

FIGURE 2.2
Example of ordinal scale.

Like nominal scales, ordinal scales have the property of equivalence—there is no relationship between numbers and choices on the scale. The numbers are used only for classification purposes. In addition, ordinal scales have a relational property that can show that an object (attribute, preference, score) is greater than or less than any other object.

Statistical analyses that require addition and subtraction cannot be performed on ordinal data; therefore, the mean and standard deviation are inappropriate calculations for ordinal scales. The frequency and percent of objects at each rank and the median rank may be computed. The median is the appropriate statistic to describe the central tendency of ranked scores because a median requires only that the same number of scores lie above and below the midpoint in the range. An ordinal scale is displayed in Figure 2.2.

Changes in individual values of scores have no effect on the median. Further analyses of data on ordinal scales can be performed using nonparametric statistical tests. These tests are discussed in later chapters.

2.6.3 Interval and Likert-Type Scales

Interval scales exhibit all the uses and properties of nominal and ordinal scales. They can be used to categorize data such as scores in the upper third, middle third, and lower third of an array of scores. Also, it is possible to use an interval scale to rank data such as the highest score, second highest, third highest, and so on to the lowest score.

A true interval scale shows a continuous distribution such that tied scores or values are virtually impossible. However, most instruments designed to measure some aspect of human behavior lack the sensitivity of true interval scales. For example, data collected on Likert-type scales that use five, seven, or nine response categories are treated as interval data. Such instruments cannot distinguish minute differences in the traits or attributes measured, and tied scores are possible. In a crude but useful form for behavioral and social sciences research, interval scales are intended to treat each item on a scale as equal in value to any other item on the same scale. This means that each item represents the same amount of knowledge, attribute, aptitude, or interest as any other item. In other words, the scores on a true interval scale treat differences in scores as equal distances between scores.

The lack of sensitivity of interval scales used in behavioral and social sciences research does not prevent researchers from using them. In fact, data for most of the research involving test scores are analyzed as interval data. Zero is not a true point on the scale; however, researchers continue to use such scales and impose zero points arbitrarily.

The appropriateness of using parametric procedures for analysis of data from Likert-type scales is questionable for several reasons. First, the assumption that Likert-type data come from a normal distribution is contradicted by the fact that data collected using Likert-type scales are bounded by responses on a continuum of values on the scale; whereas data from a normal distribution represent all real values between plus and minus infinity. Points on Likert-type scales measure respondents' extent of agreement or disagreement on a continuum for the specific items questioned.

In addition to the normality problem, other problems concern the distance between points on the scale and the difference in strength of one response when compared with any other response. The continuum for a five-point scale may be as follows: strongly agree, agree, undecided, disagree, strongly disagree. These points are designated as 5, 4, 3, 2, and 1, respectively. Researchers cannot be sure that the response of 4, for example, carries the same intensity of agreement for each respondent. We cannot say with total certainty that a response of 4 carries the same intensity of agreement from item to item for the same respondent or be totally confident that the distance between 4 and 5 is the same as the distance between 2 and 3 on the scale.

Perceptions of respondents raise questions about the equality or intensity of sameness or differences between adjacent points on a scale between respondents and also between items. Consider the following scenario. If a group of students in the eighth grade entered an art contest, one student would win first place, one second place, and one third place. The judges may rank the students' work on originality, beauty, special theme, skill, and overall appearance. The third-place winner would be in last place, unless there were more places, say fourth and fifth, behind him or her.

The extent of agreement among judges on the first- and second-place winners is difficult to ascertain. The only information we have is that the judges agreed on the first-, second-, and third-place winners. No attempt would be made to discern the difference in the extent of agreement of judges between the first and second place, the second and third place, and the third and fourth place. The problems of extent of agreement or intensity of responses and distance between points on Likert-type scales exist because the magnitude of the numbers on the scale is not comparable. Experienced researchers recognize that most scales do not have equal intervals. Many prefer to think of the assumption of *equal intervals* between points on the scale as *equal-appearing intervals*.

2.6.3.1 Scale characteristics

Interval scales allow a researcher to (1) count observations for a variable as with nominal data, (2) show greater-than and less-than relationships among numbers as with ordinal data, and (3) show ratios between equivalent sets of numbers. True interval scales provide a ratio of any two intervals on a scale that is independent of the unit of measurement and a true zero point. This means that the ratio will be the same for any two sets of two numbers on the scale. However, ratios do not provide meaningful interpretable information about the attributes being measured. The distance between any two points on the scale has meaning. For example, on a 100-point spelling test, a person can score 30 points, 70 points, or 100 points. Each item is considered to represent the same amount of spelling knowledge if all words are at the same level of difficulty. Caution should be taken when interpreting interval scales. In the case of a spelling test, words may progress in level of difficulty. The first words may be easier to spell than the last words. For example, we could not say that the level of difficulty between any 2 words among the first 20 would be as difficult as the level of difficulty between any 2 words from the last 20.

Since the zero point is arbitrary, a score of zero on a spelling test would not mean that a person has zero spelling ability. Measurement of temperature is a classic example of using an interval scale with an arbitrary zero point. Celsius and Fahrenheit scales are both used to measure temperature. Each scale uses a different metric for temperature, and each scale has a different arbitrary zero point. One can transform the temperature recorded on one scale to the other scale because the two scales are linearly related using the formula: Fahrenheit (F) degrees = 9/5 × Celsius (C) degrees + 32. The formulae for converting degrees Fahrenheit to degrees Celsius and vice versa are as follows:

$$F = 9/5 \times C + 32$$

$$C = (F - 32) \times 5/9$$

The ratio of difference is the same at equivalent points on the two scales, as illustrated in Figure 2.3.

Fahrenheit	32	50	68	86
Celsius	0	10	20	30

FIGURE 2.3
Interval scale transformations for Fahrenheit and Celsius temperatures.

An interval scale is a true quantitative measure and thus permits the computation of statistics such as the mean, variance, and standard deviation that are necessary for parametric tests such as t-tests and F-tests, which assume that data are distributed normally.

Much research in behavioral and social sciences uses interval scales, even though the intervals may be imprecise representations of the amounts of items being measured. There is considerable dispute about the appropriate use of parametric statistical tests for data collected on Likert-type scales. Questions such as the following are at the heart of the controversy:

At what point on the scale does a response of "strongly agree" blend toward a response of "agree"?

Is the intensity of agreement between "strongly agree" and "agree" the same as the intensity of agreement between "disagree" and "strongly disagree"?

Is the scaled variable normally distributed in the individuals responding to the scale?

Does a response of "strongly agree" on one item carry the same level of intensity as a "strongly agree" response on any other variable?

These kinds of questions suggest that Likert-type scales represent rank or ordinal data.

Likert-type scales present a continuum of response categories. For example, data related to participant perceptions, interests, attitudes, and opinions are collected frequently on 7- and 5-point scales. A high score represents more of the attribute being measured (unless, of course, a low score is desirable). For example, participants may be asked their attitudes toward a program, problem, or issue using a 5-point Likert-type scale where 5 = strongly agree, 4 = agree, 3 = undecided, 2 = disagree, and 1 = strongly disagree. Figure 2.4 displays a Likert-type scale.

On a 40-item scale with 5 response options, the highest possible score is 200 and the lowest possible score is 40, indicating the most positive and least positive attitudes, respectively. Some theorists who specialize in scales of measurement suggest that data for behavioral and social sciences research

---

Please circle the number that best expresses your attitude toward statistics courses.

SA = Strongly Agree; A = Agree; U = Undecided; D = Disagree; SD = Strongly Disagree

SA	A	U	D	SD
5	4	3	2	1

1. All college students should be required to take at least one statistics course.

FIGURE 2.4
Example of Likert-type scale.

collected on interval-type scales should be treated as nonparametric. Such an extreme position as proposed by Stevens more than 60 years ago does not account for the progress in measurement theory and improved instrument development since then. It is important to note that interval scales may be created successfully. Further discussion of data collected on interval scales is presented in later chapters.

2.6.4 Ratio Scales

Units of measurement on a ratio scale represent equal amounts of the attribute being measured. The ratio of any two points on a ratio scale is independent of units of measurement such as ounces or grams. For example, the difference between 40 and 50 is the same as the distance between 60 and 70. Unlike interval scales that have arbitrary zero points, ratio scales can indicate an absence. For example, zero ounces are no ounces; zero mass is no mass; and so on.

The true zero point allows researchers to claim the total absence of an item and to compare two observations as ratios or percentages. One can say that A is two or three times as much of something as B. Ratio measurement scales are used most often to measure variables such as age, weight, height, time, or money. Caution should be exercised when making ratio comparisons if variables are not measured on a ratio scale. One cannot say that a score of 100 on a test represents twice the amount of knowledge as a score of 50 or that a temperature of 32 degrees Fahrenheit is twice as cold as a temperature of 64 degrees Fahrenheit. For example, ratio comparisons cannot be made on variables such as intelligence, attitude, perceptions, or motivation. These variables have arbitrary zero points.

2.6.4.1 Scale characteristics

A ratio scale is like an interval scale but has a true zero point. All arithmetic operations can be performed on the numerical points on the scale and on the intervals between points on the scale. Ratio scales exhibit equivalence, show greater-than and less-than relationships between values, and reveal ratios between any two intervals and ratios between any two points (values) on the scale. Figure 2.5 displays two items on a ratio scale.

Most data of interest in the behavioral and social sciences do not meet the requirements for ratio scales. The unit of measurement used for ratio scales

1. Please estimate the number of words that you can type in a minute with no errors. _____
2. Please estimate the amount of time in minutes that you spend weekly watching TV. _____

FIGURE 2.5
Example of ratio scale.

may be subjective; however, the numbers on the scale represent true points or values with a true zero. Nonparametric statistics can be applied to data collected on ratio scales; however, this will seldom be the case for research in the behavioral and social sciences.

Chapter Summary

Nonparametric statistics is a broad category of statistics. Unlike parametric statistics, nonparametric statistics make no assumptions about the properties of a distribution from which data are drawn. These procedures are more suited for small sample sizes because larger samples are needed to meet the normality assumption of parametric tests. Nonparametric statistics tend to be less powerful than parametric statistics, particularly in situations where data meet the normality assumption.

Data that are amenable to nonparametric analyses are measured using ordinal or nominal scales, and sometimes interval or ratio scales; whereas parametric statistics require interval or ratio data. A major advantage of nonparametric statistics is that data taken on interval and ratio scales can be transformed to nominal or ordinal data so that nonparametric statistics can be applied. Such data transformations provide flexibility, allowing nonparametric procedures to be applied to many research problems in the behavioral and social sciences.

Parametric tests make certain assumptions about the properties of a distribution from which samples are taken. The properties of a distribution are defined by its shape (normality), midpoint (mean), and spread of scores (variance and standard deviation). These measures represent parameters of a population. Parametric statistics are robust to moderate degrees of nonnormality if the departure from normality is not too severe. Before conducting a statistical test, researchers should check that the data meet the assumptions of the test to verify whether the data are appropriate for that specific test.

Measurement requires that numbers be assigned to observations for the purpose of quantifying the observations. The observations represent the variables for research. The type of variable and the type of measurement desired dictate the manner in which the numbers are assigned. Numbers may be assigned to a personal attribute such as gender (1 = male and 2 = female) for the purpose of counting the number of men and women in a group; numbers may be assigned to test scores (95, 87, 79) for the purpose of organizing, describing, and summarizing a set of scores; numbers may be assigned for placement in a group (first, second, third) to indicate rank; or numbers may be assigned to factors (time, speed, age) to show ratios, equivalence, and relationships. The kind of scale used to collect data defines and limits the kinds of statistical procedures that may be performed on those data.

Selecting appropriate statistical tests requires consideration of the overall research design for a study. Research designs are based on criteria such as the population from which samples will be drawn and the kinds of measurement tools to be used. Other important elements of the design include research questions to be answered, hypotheses to be tested, operational definitions of the measured variables, methods and procedures for collecting the data, and power of the statistical tests applied. Research designs are tied to statistical models, and statistical models must meet certain requirements if the statistics to be applied are to yield trustworthy results. Requirements vary from one statistical procedure to another. Statistical models and the kinds of measurement tools specify the conditions for different tests. Procedures may be applied to tests whether or not a statistical model meets the required assumptions.

Assumptions for different statistical tests are usually published in statistical references. Assumptions should be verified prior to conducting a procedure, even though some parametric tests are robust enough to violate assumptions under certain conditions. A statistical test is valid only to the extent that the assumptions are met or to the extent of the kinds and severities of violations. The results of research may be inaccurate or misleading if a flawed design is used. Consequently, it behooves researchers to verify that statistical procedures are appropriate for a research design.

Before a statistical analysis can be performed, one or more quantitative or qualitative variables must be observed. The observations may be expressed as nominal, ordinal, interval, or ratio data. Observations are expressed as numbers that represent specific qualities of the data such as a category of response, magnitude of an attribute or property, distance between intervals, or an absolute or arbitrary zero point. Numbers are assigned to each data type (nominal, ordinal, interval, and ratio) according to a set of rules.

The remaining chapters in this book present a discussion of nonparametric procedures used frequently in behavioral and social sciences research. Each chapter is organized around a group of statistical procedures. Specific tests will be presented within each group. Examples are incorporated for each procedure, and formulas are included to demonstrate manipulation of the data. Each chapter shows the IBM-SPSS procedures used to calculate the statistics. Chapter summaries and exercises are intended to enhance student learning.

Student Exercises

2.1 Locate five research articles on topics of interest for your field of study and respond to the following for each article:

    a. Record the author(s), article title, journal name, volume and issue number, and page numbers.

    b. Write a statement of the problem investigated.

      c. List the kinds of measurement instruments used to collect the data.

      d. List the statistical tests used to analyze the data.

      e. Which, if any, other kinds of instruments could have been used to address the problem investigated?

2.2 Which type(s) of measurement instrument(s) is (are) most appropriate for your field of study? Why?

2.3 Select two topics of interest in your field of study, and complete the following for each topic. The topics may be two of the five that you used to complete Exercise 2.1. Develop two measurement scale items appropriate for each of the two topics you selected for

      a. nominal scale,

      b. ordinal scale,

      c. interval scale,

      d. ratio scale.

2.4 List potential advantages of using nonparametric statistics in your field of study.

2.5 List potential pitfalls of each type of measurement scale (nominal, ordinal, interval, and ratio). For example, a potential pitfall of using a nominal scale is that all possible categories may not be included.

References

Aerts, M., Claeskens, G., and Hart, J. D. (1999). Testing the fit of a parametric function. *Journal of the American Statistical Association*, 94, 869–879. http://www.jstor.org/stable/2670002

Andersen, P. K. and Ronn, B. B. (1995). A nonparametric test for comparing two samples where all observations are either left- or right-censored. *Biometrics*, 51(1), 323-329. http://www.jstor.org/stable/2533338

Black, T. R. (2009). *Doing Quantitative Research in the Social Sciences: An Integrated Approach to Research Design, Measurement, and Statistics.* Thousand Oaks, CA: Sage.

Gibbons, J. D. and Chakraborti, S. (2010). *Nonparametric Statistical Inference*, 5th ed. Boca Raton, FL: Chapman & Hall/CRC.

Hollander, M. and Wolfe, D. A. (1999). *Nonparametric Statistical Methods*, 2nd ed. New York: John Wiley & Sons.

Sirkin, R. M. (2006). *Statistics for the Social Sciences*, 3rd ed. Thousand Oaks, CA: Sage.

Tuckman, B. W. and Harper, B. E. (2012). *Conducting Educational Research*, 6th ed. New York: Rowman & Littlefield.

Yohal, V. J. and Zamar, R. H. (2004). Robust nonparametric inference for the median. *Annals of Statistics*, 32, 1841–1857. http://www.jstor.org/stable/3448557

3

Analysis of Data to Determine Association and Agreement

The nonparametric procedures presented in this chapter allow researchers to test for association, independence, or agreement among two or more variables. Association between two variables is a broader term for a relationship between two variables. The term *relationship* is used with parametric data to show a linear relationship between two continuous or ratio variables. The term *association* is used generally to show any relationship. For all practical purposes, there is no important difference between the two terms, and they are often used interchangeably.

If you have ever taken a basic course in parametric statistics, you probably recall that relationships between and among variables are expressed in terms of strength, direction, and statistical significance. As in parametric tests, the values for association of two or more nominal or ordinal scaled variables can range from –1 to +1. A negative coefficient indicates that variables are changing in the opposite direction, and a positive coefficient means that variables are changing in the same direction. The closer the coefficient is to zero, the weaker is the association. A coefficient of zero means absolutely no association between the variables. An important concept is that the association does not mean causation or dependency of one variable on another.

The range of values for measures of agreement varies with the type of test conducted. For example, a measure of agreement can range from zero indicating absolutely no agreement to 1 indicating perfect agreement. Some tests that ordinarily produce positive levels of agreement can yield negative coefficients; however, they are exceptional and will be discussed in a later section of this chapter.

Some measures of agreement range from ±1 as the coefficient for association. As with interpretation of the strength of an association, interpretation of the extent of agreement from very weak to very strong is somewhat subjective. The subjectivity is typically based on the type of research conducted, the context of the study, consequences of the study, and the theoretical and/or empirical basis for the study. Conventional wisdom usually accepts coefficients of 0.20, 0.60, and 0.80 as indications of weak, moderate, and strong levels of agreement, respectively.

This chapter covers the most commonly used nonparametric measures of association and agreement. Some of these tests are closely related, as you will see. The chapter begins with a discussion of the chi-square test of association and independence. Next, the contingency coefficient, phi coefficient, and Cramer's coefficient *V* are covered. Information about Kendall's tau_b and Kendall's tau_c is included. This chapter also presents a discussion of the kappa statistic and the Spearman rank-order correlation. Each test is discussed along with an example that illustrates its applications. Formulas for each test and calculations using a hand calculator are included simply to demonstrate the ways the data are manipulated and to help one conceptualize the calculations.

The SPSS procedures are included to show steps in the conduct of the analysis for each test, and each test is illustrated with an example. The chapter ends with a summary and student exercises to reinforce learning and to practice the tests presented.

The data sets for the examples and exercises of this chapter can be retrieved online from the file Chapter3data.zip under the Downloads tab at http://www.crcpress.com/product/isbn/9781466507609.

3.1 Pearson Chi-Square Test of Association and Independence

The Pearson chi-square test of association and independence is a popular technique for evaluating the association or independence of two or more variables. The symbol for chi-square is written as χ^2. The term *chi-square* and the symbol χ^2 are used interchangeably. As a reminder, *chi* is pronounced to rhyme with *pie* and not to rhyme with *gee*.

The purpose of the chi-square test is to determine statistically significant differences between frequencies of responses on discrete variables for two independent groups. The test assesses whether the observations or responses on one variable are associated with or are independent of the observations or responses on the other variable. Responses are mutually exclusive, and classification of cases into groups is also exclusive.

The data for the chi-square test can be dichotomous, nominal, ordinal, or grouped into intervals. If the data are grouped into intervals, they should be in the form of numerals, not percents or ratios. Although the Pearson chi-square test of association and independence can be applied to ordinal data, the Spearman rho procedures are preferable for ordinal data. The Spearman rho procedures are presented in Section 3.6.

The chi-square test of association and independence requires that responses for two independent groups be compared on a dependent variable with two possible response categories in a 2 × 2 design. For example, a study using nominal data such as gender with two categories (male and female) and two religious preferences (catholic or protestant) would assess whether religious preference was independent of gender. Data for a randomly selected sample of males and females could be recorded in a two-by-two (2 × 2) contingency table for ease of calculation. The design was chosen because the study utilizes two variables (gender and religious preference), each with two categories.

A 2 × 2 contingency table has four cells in which the frequency counts for each occurrence are recorded. The 2 × 2 contingency table is the simplest type when multiple rows and colums are involved. Usually groups are identified on the columns and variable responses are recorded on the rows. However, this arrangement is not mandatory; the rows and columns may be interchanged so that the variable of interest (dependent variable) is reported on the columns and groups are recorded on the rows.

When two or more groups are compared on a specific variable with more than two response categories, such as choice of a political party (democratic, republican, or independent), the chi-square test of association and independence is called an $r \times c$ (row by column) design. The null hypothesis for the chi-square test of independence is that there are no group differences with respect to group responses on a categorical variable. Also, the null hypothesis may be stated as some behavior or preference is independent of a specific attribute of the respondents. For example, the null hypothesis may be that the preference for a certain kind of food (pizza, steak, macaroni and cheese) is independent of gender. Stated another way, the null hypothesis may be worded as "no statistically significant difference between boys and girls on their preference for favorite food." The null hypothesis is written in the following form:

H_0: No difference in favorite food based on gender.

The alternative hypothesis is written as follows:

H_A: Preference for favorite food varies by gender.

Small differences will result in a small chi-square value when observed and expected frequencies are close in number. In such cases, the null hypothesis cannot be rejected but must be retained or supported. Remember never to accept a null hypothesis! Theoretically, one can retain or support a null hypothesis, but not accept it. This idea is related to the central limit theorem. Conversely, when the number of observations from one group differs from those of another group by chance alone and not an independent variable in the study, the chi-square value will be large. A large chi-square value yields a statistically significant p-value, which will lead to rejection of the null hypothesis.

The chi-square test is based on differences in the numbers of observed and expected frequency of responses for all combinations of independent and dependent variables. The research hypothesis underlying the chi-square test is that there is no difference between the observed and expected frequency of the number of observations on one variable based on the frequency of the number of observations for a second variable. In other words, the two variables are not independent of one another. If the observed responses on the dependent variable for one group are not statistically different from those in a second group, then one cannot claim a group difference. If the observed responses exceed the responses expected by chance alone, then one can conclude that statistically significant differences exist between the two groups.

Data for the chi-square test of association and independence should meet a few assumptions before one can be assured that the test will yield trustworthy results. The assumptions can be met through the design of the study. For example, an adequate sample size is important so that the sample is representative of the population from which the sample was selected. Representativeness allows the researcher to generalize findings to the broader population. The population from which the sample is taken need not be distributed normally; however, if inferences are to be made, a population at least 10 times larger than the sample is desirable. In addition, the sample should be randomly selected from a clearly identified population.

The minimum individual cell count should be 5 for small samples, with no more than 20% of the cells having an expected cell count less than 5 for large samples. No cells should have zero counts for samples of any size. For tables larger than 2 × 2 with cell counts less than 5, the researcher may be able to pool cells (groups or variables) with low counts. Caution should be taken to ensure that combining cells does not result in a loss of data that diminishes interpretation of the results.

Responses for any specific observation should be independent of any other responses for the same observation. Each response or observation can be represented in only one cell. In other words, no observation can be represented by more than one level of each variable. Specifically, if a question asks for a person's religious preference (catholic or protestant), an individual can select catholic or protestant but not both.

The chi-square statistic can be determined with a hand calculator but such calculations are undesirable. Modern statistical software makes manual calculations, which can waste valuable research time, unnecessary. In addition to inefficiency, manual calculations are prone to error. Hand calculations are shown here only for illustration purposes for those who wish to understand the details of the calculation of the chi-square statistic.

The following example shows the calculations for the chi-square test of independence. Later, you will see that computer software such as SPSS can return quick and accurate results with only a few clicks on the menus.

Example 3.1

A development officer for athletics is interested in whether gender is related to preferences for certain sports. The officer is interested only in an individual's preference for baseball or football. The first step is formulating a hypothesis to test. The null hypothesis (H_0) for this example tests whether the association between sport preference and gender is statistically significant. The null hypothesis is stated as:

H_0: Sport preference is independent of gender.

The alternative hypothesis may be written as:

H_A: Sport preference is not independent of gender.

Next, the development officer selects a random sample of 109 males and females and asks each of them to indicate his or her preference for baseball or football. Respondents can select only one preference. Expected frequencies can be obtained by multiplying the total number in a row by the total number in a column and dividing the product by the total number in the sample. This procedure should be repeated by row for each column.

Row 1 total is multiplied by column 1 total, and row 2 total is multiplied by column 2 total. The chi-square statistic is the sum of the differences between the observed and expected frequencies divided by the expected frequencies for each cell in the table. This simple formula may be helpful for remembering the calculation of expected frequencies

$$E_{ij} = \frac{R_i C_j}{N}$$

where E_{ij} is the frequency for every cell, R_i is every row, C_j is every column, and N is the sample size. The expected frequency in each cell (column and row intersection) will be proportionate to the row and column totals if responses are independent of group membership. Responses for the sport preference by gender study are shown in Table 3.1.

The formula for the chi-square test of association and independence is given in Figure 3.1. The following formulae illustrate the application of the chi-square formula to the data in Table 3.1. Expected values are calculated as (row × column)/total N for each row by each column. The expected frequency calculations are:

$$R_1, C_1 : \frac{57 \times 50}{109} = 26.15$$

$$R_1, C_2 : \frac{57 \times 59}{109} = 30.85$$

$$R_2, C_1 : \frac{52 \times 50}{109} = 23.85$$

$$R_2, C_2 : \frac{52 \times 59}{109} = 28.15$$

TABLE 3.1

Observed Values for Sport Preference by Gender

	Preference		
Gender	Baseball	Football	Row Total
Male	17	40	57
Female	33	19	52
Total	50	59	109

$$\chi^2 = \sum_{i=1} \sum_{j=1} \frac{\left(n_{ij} - E_{ij}\right)^2}{E_{ij}}$$

FIGURE 3.1
Formula to calculate chi-square statistic where n_{ij} is the frequency of observations for a specific row for each column. E_{ij} is the expected number of observations in a specific row for each column under a true null hypothesis.

χ^2 equals the sum of observed frequencies minus expected frequencies squared and divided by the expected frequencies. The following calculations demonstrate the differences between the observed and expected frequencies squared divided by the expected frequencies:

$$\chi^2 = \frac{(17 - 26.15)^2}{26.15} + \frac{(40 - 30.85)^2}{30.85} + \frac{(33 - 23.85)^2}{23.85} + \frac{(19 - 28.15)^2}{28.15}$$

$$= 3.20 + 2.71 + 3.51 + 2.97 = 12.39$$

Results of the chi-square test are based on a chi-square distribution that has a range between zero and one—not to be confused with the value of the chi-square statistic. Chi-square cannot have a negative value. The degrees of freedom (dfs) are important parameters for a chi-square distribution. Each time we select a random variable from a standard normal population, the degrees of freedom increase. As the degrees of freedom increase, the area under the normal curve increases and the chi-square distribution approaches a normal distribution.

The degrees of freedom for a chi-square distribution are calculated by the number of rows (r) in a table minus 1 multiplied by the number of columns (c) in a table minus 1 [df = $(r - 1)(c - 1)$]. As you will learn later, some of the other nonparametric tests are also based on a chi-square distribution.

Interpretation of the chi-square results is based on the value of the chi-square statistic at the appropriate df. The chi-square statistic is distributed asymptotically as an approximate chi-square distribution with df equal to $(r - 1)(c - 1)$. The asymptotic nature of the chi-square distribution is depicted in Figure 3.1.

The significance of the chi-square value can be identified in two ways. First, if the observed probability (alpha) level is less than the a priori probability level (usually < 0.05), we must reject the null hypothesis of independence and conclude that the two variables are associated or dependent on one another. Second, we may compare the obtained chi-square value to the critical value on a chi-square table at the appropriate df to ascertain whether an obtained chi-square value is greater than the critical value. If the obtained value is greater than the critical value in the table, then the chi-square statistic is significant, and we can claim a statistically significant association or dependency between the two variables. Consequently, we must reject the null hypothesis of independence.

If the observed chi-square value in our example of sport preference by gender (12.39) were compared to a table of chi-square critical values, we would see that the obtained chi-square value exceeds the critical value of 3.841 at 1 df for alpha = 0.05, indicating statistical significance. In fact, the calculated chi-square value of 12.39 exceeds the table value of 10.828 at alpha = 0.001. Tables of chi-square critical values can be found in most basic statistics books and on the Internet. The development officer can confidently reject the null hypothesis and declare that a preference for football or baseball is dependent on gender. In other words, one can conclude that sport preference and gender are related.

The chi-square statistic can be calculated easily using the SPSS software. Data may be entered into a program spreadsheet as raw data or as weight data (weight cases). A partial data view of the SPSS spreadsheet using raw data for Example 3.1 is shown in Figure 3.2. The weight cases method is effective for large sample sizes. Figure 3.3 shows the data view using this method.

As you can see from Figure 3.3, the weight cases method of entering data is very efficient. One simply has to name the two variables [gender

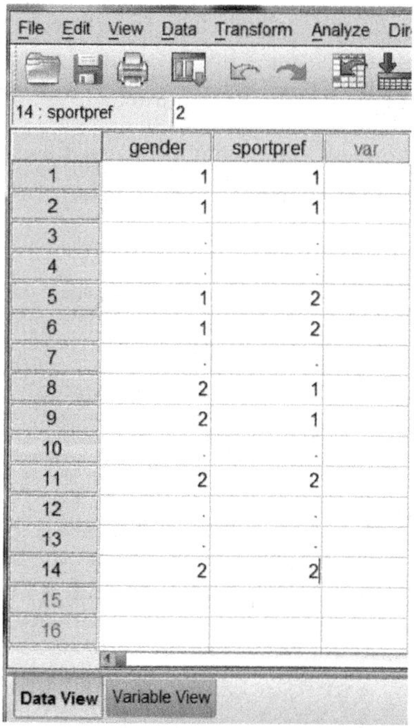

FIGURE 3.2
Partial data view for sport preference by gender using raw data.

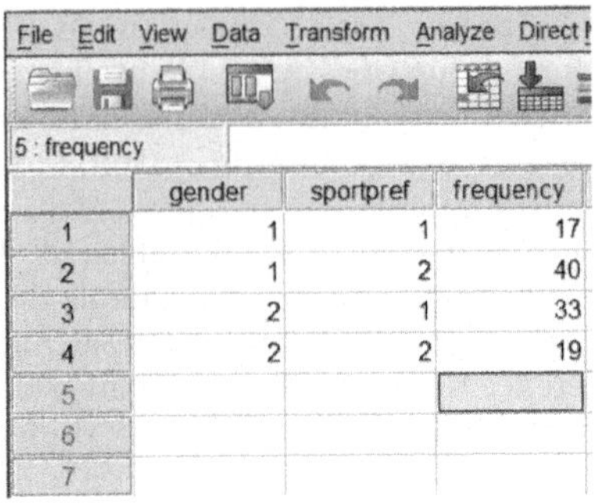

FIGURE 3.3
Weighted case data for sport preference by gender.

and sport preference (sportpref) in this case] and create a third variable (frequency in Figure 3.3) so that the number of cases for one variable by the other can be specified. A contingency ($r \times c$) table is a good way to organize the data before transferring the frequencies to the SPSS spreadsheet.

Use the following procedure to weight the cases for the data in Table 3.1 for the sport preference by gender study.

> Click Data – Click Weight Cases – The Weight Cases dialog box will open – Select Weight cases by – Move the frequency variable to the box for Frequency Variable – Click OK.

The open Weight Cases dialog box is shown in Figure 3.4. After cases are weighted, use the following path to conduct the chi-square procedure for association.

> Analyze – Descriptive Statistics – Crosstabs – Move gender to the Row(s) box – Move sportpref to the Column(s) box – Click Display clustered bar charts in the lower left section – Click Statistics – Chi-square – Continue – Click Cells – (Observed should already be selected) – Click Expected – Click Row in the Percentages section – Continue – Click Exact – Click Exact within the Exact Tests dialog box – Click Continue – Click OK.

Figure 3.5 shows the SPSS path to access the Crosstabs function, and Figure 3.6 shows the open Crosstabs dialog box.

Figure 3.7 displays the Crosstabs Statistics dialog box. Figures 3.8 and 3.9 show the Crosstabs Cell Display dialog box and the Exact Test dialog box, respectively.

As shown in Figure 3.10, the results for the chi-square test of association between sport preference and gender are statistically significant: $\chi^2(1, N = 109) = 12.40, p < 0.01$.

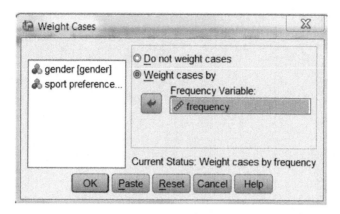

FIGURE 3.4
SPSS dialog box for weight cases.

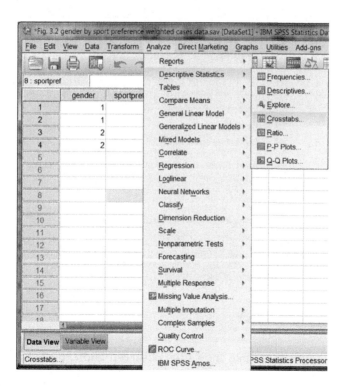

FIGURE 3.5
SPSS path to access Crosstabs.

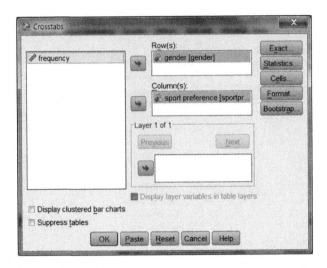

FIGURE 3.6
SPSS open Crosstabs dialog box.

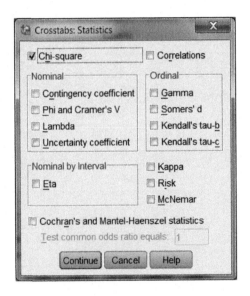

FIGURE 3.7
SPSS Crosstabs Statistics dialog box.

The cross tabulation table shows the observed and expected frequencies for sport preference by gender. Forty males (70% of all males, $N = 57$) compared to 19 females (37% of all females, $N = 52$) preferred football to baseball. Observed and expected frequencies for the sport preference by gender example are shown in Figure 3.11.

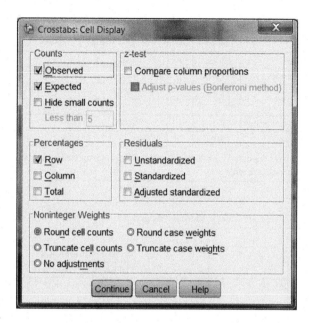

FIGURE 3.8
SPSS Crosstabs Cell Display dialog box with selected counts and percentages.

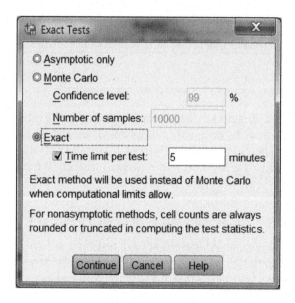

FIGURE 3.9
SPSS Exact Test dialog box with Exact selected.

	Value	df	Two-Sided Asymptotic Sig.	Two-Sided Exact Sig.	One-Sided Exact Sig.
Pearson chi-square	12.391[a]	1	0.000		
Continuity correction[b]	11.074	1	0.001		
Likelihood ratio	12.623	1	0.000		
Fisher's exact test				0.001	0.000
Linear-by-linear association	12.278	1	0.000		
Number of valid cases	109				

[a] No cells (0.0%) have expected counts less than 5; minimum expected count is 23.85.
[b] Computed only for 2 × 2 table.

FIGURE 3.10
Results of chi-square test for sport preference by gender.

			Sport Preference		Total
			Baseball	Football	
Gender	Male	Count	17	40	57
		Expected Count	26.1	30.9	57.0
		% within gender	29.8%	70.2%	100.0%
	Female	Count	33	19	52
		Expected Count	23.9	28.1	52.0
		% within gender	63.5%	36.5%	100.0%
Total		Count	50	59	109
		Expected Count	50.0	59.0	109.0
		% within gender	45.9%	54.1%	100.0%

FIGURE 3.11
Observed and expected frequency for sport preference by gender.

Bar charts are appropriate for presenting a quick visualization of nominal data. They provide an easily understood way to show comparisons among groups. The clustered bar chart in Figure 3.12 shows a larger number of males preferring football and a larger number of females preferring baseball.

3.1.1 Contingency Tables with More Than Two Rows or Columns

Thus far, we have discussed the chi-square test for 2 × 2 contingency tables. The chi-square statistic can be applied to larger contingency tables as well. A statistically significant chi-square value for contingency tables with more than two rows and/or columns requires the performance of several different chi-square tests to determine where the differences are. Example 3.2 illustrates a 3 × 2 contingency table for observed and expected frequencies for three possible responses on a single variable (preferred method of conflict resolution) for two groups (gender).

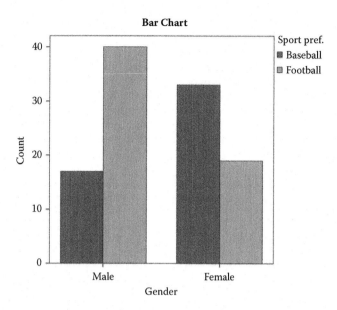

FIGURE 3.12
Bar chart for sport preference by gender.

TABLE 3.2

Observed and Expected Frequencies (in Parentheses)
for Conflict Resolution by Gender

Preferred Method	Men	Women	Combined
Mediator	14 (21.3)	31 (23.7)	45
Self Negotiating	21 (18.4)	18 (20.6)	39
Ignoring	17 (12.3)	9 (13.7)	26
Total	52	58	110

Example 3.2

A marriage counselor is interested in whether men or women are more likely to seek professional counseling for conflicts within their marriages, to ignore conflicts, or to try to resolve conflicts themselves. The counselor collects data from a sample of 110 men and women and asks the participants to answer questions about conflict resolution in a manner that reflects their true feelings. The 3 × 2 contingency table is displayed in Table 3.2.

For this example, the Pearson chi-square is significant beyond the 0.05 alpha level of significance [$\chi^2(2, N = 110) = 8.81, p = 0.01$, exact p-value = 0.01]. Therefore, the null hypothesis of no association between the groups for choice of method of conflict resolution is rejected at the 0.05 level.

The conclusion is that choice of method for conflict resolution is associated with group membership (gender). Figure 3.13 shows results for the chi-square test. Figure 3.14 displays a bar chart for the three methods of conflict resolution by gender.

As shown in Figure 3.13, the omnibus test (test for hypothesis) is statistically significant; however, we do not know from the chi-square value or the p-value alone which methods of conflict resolution are independent of gender because the measured variable has more than two categories. Associations may be found for some categories, but not for others. The conduct of several follow-up chi-square analyses is necessary to discover the levels of one variable that are significantly associated with levels of the other variable.

	Value	df	Two-Sided Asymptotic Sig.	Two-Sided Exact Sig.	One-Sided Exact Sig.	Point Probability
Pearson chi-square	8.813[a]	2	0.012	0.013		
Likelihood ratio	8.990	2	0.011	0.013		
Fisher's exact rest	8.783			0.013		
Linear-by-linear association	8.428[b]	1	0.004	0.005	0.003	0.001
Number of valid cases	110					
[a] No cells (0.0%) produced expected count less than 5; minimum expected count is 12.29.						
[b] Standardized statistic is −2.903.						

FIGURE 3.13
Chi-square results for conflict resolution by gender.

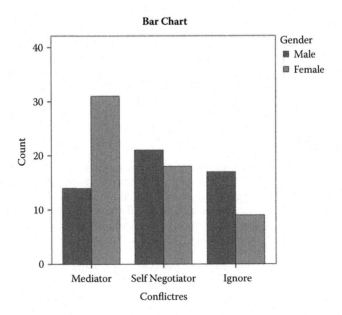

FIGURE 3.14
Bar chart for conflict resolution by gender.

Follow-up procedures can be conducted in two different ways. If we wish to portion the degrees of freedom in tables with multiple rows and two groups, the procedure requires the construction of as many 2 × 2 subtables as there are degrees of freedom in the original contingency table. The example of choice of method for conflict resolution (3 × 2 contingency table) has two degrees of freedom [$(r − 1)$ and $(c − 1)$]. Consequently, we can form two 2 × 2 subtables with one degree of freedom each.

The order of the partitions should be decided a priori. First, we construct a contingency table for differences between two levels of one variable for two categories of another variable. If the chi-square test is statistically significant, we will construct a contingency table for the next most likely difference (or comparison of most interest). If the chi-square test is not statistically significant, it is acceptable to combine the frequency for one variable (e.g., frequency of men and women combined) on both nonstatistically significant variables and let the combination of frequencies serve as the first row in the analysis of the differences for the second subtable. However, one should exercise caution when combining several levels of variables because the practice may result in a loss of valuable information.

The second way to conduct follow-up procedures for a statistically significant omnibus test requires the construction of all possible subtables in the original contingency table. A statistical technique to control for the Type I error rate is recommended for contingency tables with more than two rows or columns. For the conflict resolution data, we need to develop three 2 × 2 contingency tables. The first contingency table could test for differences between the mediator and self negotiator choices for men and women. The SPSS path to conduct pairwise follow-up tests is as follows:

> Click Data – Select Cases – If condition is satisfied – Click If – Type in conflictres = 1 or conflictres = 2 to identify the cases – Click Continue – OK.

If your data are entered using the weight cases method, remember to invoke the Weight Cases function presented in Section 3.1 prior to your analysis. Now the data are ready for the chi-square test of independence of men and women on their choices of methods for conflict resolution. Refer to earlier in this section for the SPSS path to conduct follow-up comparisons. Repeat the chi-square procedures, selecting different combinations of variables each time a different comparison is made.

The chi-square test results between the mediator and self negotiator methods of conflict resolution show a statistically significant association between method of conflict resolution and gender [$\chi^2(1, N = 84) = 4.44, p = 0.04$]. Even though the *p*-value is 0.04, one should exercise caution in claiming statistical significance because multiple comparisons are made. Figure 3.15 shows the chi-square test results.

More men (21 or 60%) than women (18 or 37%) preferred the self negotiator method of conflict resolution. Conversely, more women (31 or 63%) than

	Value	df	Two-Sided Asymptotic Sig.	Two-Sided Exact Sig.	One-Sided Exact Sig.	Point Probability
Pearson chi-square	4.443[a]	1	0.035	0.046	0.029	
Continuity correction[b]	3.557	1	0.059			
Likelihood ratio	4.471	1	0.034	0.046	0.029	
Fisher's exact test				0.046	0.029	
Linear-by-linear association	4.390[c]	1	0.036	0.046	0.029	0.020
Number of valid cases	84					

[a] No cells (0.0%) had expected count less than 5; minimum expected count is 16.25.
[b] Computed only for 2 x 2 table.
[c] The standardized statistic is −2.095.

FIGURE 3.15
Results of chi-square test for mediator and self negotiator for conflict resolution by gender.

			Conflict Resolution		
			Mediator	Self Negotiate	Total
Gender	Male	Count	14	21	35
		Expected Count	18.8	16.3	35.0
		% within gender	40.0%	60.0%	100.0%
	Female	Count	31	18	49
		Expected Count	26.3	22.8	49.0
		% within gender	63.3%	36.7%	100.0%
Total		Count	45	39	84
		Expected Count	45.0	39.0	84.0
		% within gender	53.6%	46.4%	100.0%

FIGURE 3.16
Observed and expected frequencies for mediator and self negotiator conflict resolution by gender.

men (14 or 40%) preferred a mediator for conflict resolution. Observed and expected frequencies for men and women on the mediator and self negotiator methods of conflict resolution are reported in Figure 3.16.

Results of the remaining two comparisons are given here; however, to conserve space, their accompanying SPSS outputs are not included. Results for preferences of men and women in resolving conflicts by engaging a mediator or ignoring conflicts were significant [$\chi^2(1, N = 71) = 7.87, p < 0.01$]. More women (31 or 78%) preferred a mediator than men (14 or 45%); 17 or 55% of the men preferred to ignore conflict, whereas only 9 or 23% of the women preferred that method of conflict resolution.

The third contingency table compared the conflict resolution methods of self negotiation and ignoring the conflict between men and women. Results of this comparison were nonsignificant [$\chi^2(1, N = 65) = 0.86$, $p = 0.36$, exact p-value = 0.44]. Clearly, there is no relationship between preference of conflict resolution and gender for the self negotiation and ignoring methods.

Since we conducted three follow-up comparisons, it is appropriate to control for family-wise error rate across the comparisons. You may recall that Holm's Bonferroni procedure is an acceptable adjustment made by dividing the alpha level by the number of comparisons to set a more stringent alpha level for the first comparison. Next, conduct all the follow-up tests and then start with the smallest p-value to see whether the new alpha level is exceeded. If it is, stop the comparisons. If not, divide the alpha again by $n - 1$ of the remaining comparisons and compare the new alpha with the next smallest p-value.

Each time a comparison is made, the original alpha is divided by the remaining number of comparisons to be made. For our example, the smallest p-value of 0.005 certainly meets the criterion of significance based on 0.05/3 = 0.017. The next smallest p-value of 0.04 compared to an adjusted alpha set at 0.025 (0.05/2) is not statistically significant. At this point, we have no need to continue the comparisons.

When reporting statistical results, one should include the chi-square value and the associated asymptotic and exact p-values. Exact p-values are reported for 2 × 2 tables when exact tests (also known as Fisher's exact tests) are selected. No confidence interval is displayed because the exact p-value is accurate. A table displaying the observed chi-square statistics and p-values is appropriate when follow-up tests are made and a table showing observed and expected frequencies is recommended. A researcher should explain the procedures that were performed to control for a Type I error rate when multiple comparisons are made. These adjustments should be shown in the table. In addition, a chart is always helpful, especially for conference presentations. Some researchers choose to report contingency coefficients because the chi-square statistic does not indicate the strength of agreement. Contingency coefficients are discussed in Section 3.2.

3.2 Contingency Coefficient

The contingency coefficient is a measure of strength of association between two variables and may be called the contingency coefficient C. The coefficient is based on the Pearson chi-square statistic but it is uniquely different and thus warrants a separate discussion. The chi-square statistic tests the association and independence of observations in a 2 × 2 contingency table.

The contingency coefficient is not limited to 2 × 2 tables. The coefficient can test for independence of observations in tables with more than two rows and/or columns.

Data for a contingency coefficient should be nominally scaled. The coefficient is scale invariant in that it does not change as sample size changes so long as cell sizes change relative to each other. Since the contingency coefficient is derived from the Pearson chi-square statistic, it is easily computed by dividing the value of the chi-square statistic by the sum of the chi-square value and the number in the sample and then determining the square root of the quotient.

The test will yield an index between 0 and 1, although a coefficient of 0 or 1 would be highly unusual. Zero indicates absolutely no association between variables (total independence) and 1 indicates total dependence of one variable on the other. In other words, the closer the coefficient is to 0, the weaker the association; and the closer the coefficient is to 1, the stronger the association.

As an illustration, let us use the sport preference by gender example in Example 3.1 to compute the contingency coefficient. Recall that 109 individuals (57 males and 52 females) were asked whether they preferred baseball or football. Refer to Figure 3.10 to see that $\chi^2(1, N = 109) = 12.391$, $p < 0.01$. Taking the square root of the chi-square value divided by the sum of the chi-square value and N, we find that the contingency coefficient is 0.32 ($\sqrt{12.391/(12.391+109)}$). Interpretation of the contingency coefficient is subjective. However, one would usually consider a coefficient of 0.32 as indicating a fairly strong association between two variables.

Obtaining the contingency coefficient in SPSS follows the same menu and path as the Pearson chi-square statistic presented in Example 3.1. The only difference in the SPSS choices is that Contingency Coefficient should be selected in the Crosstabs Statistics dialog box. (See Figure 3.7.) Figure 3.17 displays the results for the contingency coefficient of the sport preference by gender study.

Two other measures of association, the phi coefficient and Cramer's V, are used to show the strength of relationship for a statistically significant chi-square value. These measures are presented in Section 3.3.

Symmetric Measure			
	Value	Approximate Sig.	Exact Sig.
Nominal by Nominal Contingency Coefficient	0.319	0.000	0.001
Number of valid cases	109		

FIGURE 3.17
Results for contingency coefficient for sport preference by gender.

3.3 Phi Coefficient and Cramer's *V* Coefficient

The close relationship of the phi and the Cramer's *V* coefficients merits their discussion in the same section. Both tests are measures of association derived from the chi-square test and both are contingency coefficients. Essentially, they are follow-ups to a statistically significant chi-square test and are often reported along with the chi-square statistic and associated probability level.

If a chi-square statistic is not statistically significant, follow-up tests of the association are meaningless. Phi and Cramer's *V* coefficients produce identical results for 2 × 2 contingency tables; however, both can assess tables with more than two rows and/or columns. The primary difference between the two techniques is that the phi coefficient indicates the direction of the association (negative or positive) between two variables and ranges from –1 to +1. However, the direction of the association is not interpretable when phi is calculated from nominal data. Cramer's *V* coefficient ranges from 0 to 1. If the direction of relationship is of interest, then phi should be selected over Cramer's *V*. Both are symmetrical measures. Variables on the rows and columns may be interchanged with no loss in interpretation of the results.

Recall that a binary (dichotomous) variable is a discrete random variable that can have only two possible outcomes, for example, success or failure, pass or fail, married or not married, graduate student or undergraduate student, etc. These dichotomous outcomes are usually coded as 0 for one outcome and 1 for the other. The codes are meaningless in terms of value.

Data for the phi coefficient and Cramer's *V* coefficient are entered into a 2 × 2 contingency table for ease of computation. For the phi coefficient, the product of the diagonal cells (upper left to lower right) is compared to the product of the off-diagonal cells (upper right to lower left); this makes it easier to see whether an association is positive or negative. Phi can reach maximum and minimum values of ±1 if the frequency of both variables is evenly split in a 2 × 2 table.

If most of the data points fall diagonally on the table, the two variables are positively associated. A negative association results when most of the data points are in the off-diagonal cells. The greater the number of frequencies on the left-to-right diagonal and the lesser the number of frequencies on the right-to-left off-diagonal, the larger will be the phi coefficient in a positive direction. Conversely, the smaller the number of frequencies on the left-to-right diagonal and the larger the number on the right-to-left off-diagonal, the larger will be the phi coefficient in a negative direction. This is so because the phi coefficient is based on a comparison of the product of the diagonal cells minus the product of the off-diagonal cells and divided by the square root of the product of the row and column margin totals.

$$\varphi = \frac{(a \times d) - (b \times c)}{\sqrt{\text{product of marginal totals}}}$$

The marginal totals are the totals in each row and the totals in each column. A simpler formula to calculate phi is to divide the chi-square statistic by the number in the sample and take the square root of the quotient ($\varphi = \sqrt{\chi^2 / n}$).

Cramer's V coefficient is another measure to assess dependence of two variables. Cramer's V is scale invariant—the degree of dependence does not change when values of the observations change relative to one another. Cramer's V coefficient is easily calculated by dividing the chi-square statistic by the product of the sample size (n) multiplied by the smaller of either the number of categories for a row (r) or a column (c) minus 1 and then taking the square root of the quotient:

$$\text{Cramer's } V = \sqrt{\frac{\chi^2}{n([r \text{ or } c]-1)}}$$

The coefficient ranges from 0 to 1, with 0 indicating absolutely no association and 1 indicating a perfect association. A general rule of thumb for interpreting both the phi and the Cramer's V is that coefficients > 0.70 indicate a strong association. Coefficients ranging from 0.30 to 0.70 show weak to fairly strong associations. A coefficient < 0.30 shows a very weak to negligible association. Example 3.3 illustrates the phi and Cramer's V coefficients.

Example 3.3

Consider a study in which campus planners are interested in the number of students who drive their cars to school or use the campus transit system. They also want to know whether more males or females drive their cars or take the campus transit system. The planners want to know the extent of the correlation between gender and driving to school and gender and using the campus transit system. To answer this question, 180 students were randomly selected for the study. They were asked whether they drive their own cars or use the campus transit system. The null hypothesis for this study is stated as follows:

H_0: Mode of transportation is independent of gender.

The alternative hypothesis is stated as follows:

H_A: Mode of transportation is dependent upon gender.

The frequency of responses from the 180 students is recorded in a contingency table (Table 3.3). The steps for conducting the analyses for phi and Cramer's V are the same as those for the chi-square test in Example 3.1, with the exception of selecting Phi and Cramer's V in the Crosstabs Statistics box. One could also select the row and column percentages and the total percentages if those data were of interest. The SPSS steps of the procedure using the Weight Cases method are repeated here for convenience.

Click Data – Click Weight Cases – The Weight Cases dialog box will open – Select Weight cases by – Move the frequency variable to the box

TABLE 3.3

Frequency of Responses by Gender for Transportation Study

Gender	Transportation		Total
	Personal Vehicle	**Campus Transit**	**Total**
Female	14	66	80
Male	89	11	100
Total	103	77	180

for Frequency Variable – Click OK. Analyze – Descriptive Statistics – Crosstabs – Move gender to the Row(s) box – Move transportation to the Column(s) box – Click Statistics – Chi-square – Click Phi and Cramer's *V* – Continue – Click Cells – (Observed should already be selected) – Click Expected – Click Row in the Percentages section – Click Continue – Click Exact – Click Exact within the Exact Tests dialog box – Click Display clustered bar charts – Click OK.

Results of the test are statistically significant [$\chi^2(1, N = 180) = 92.82$, $p < 0.01$]. Therefore, we must reject the null hypothesis of independence and declare that the mode of transportation is dependent upon gender. The phi and Cramer's *V* statistics (0.72) indicate a very strong association between mode of transportation and gender. Figure 3.18 shows the chi-square values, levels of significance, and the phi and Cramer's *V* coefficients.

Sixty-six (83%) of the females use the campus transit system, and only 11 (11%) of the males use it. Conversely, only 14 (18%) of the females use their personal vehicles, whereas 89 (89%) of the males use their personal vehicles. Figure 3.19 displays the frequency and percent of mode of transportation by gender.

As stated previously, for a 2 × 2 contingency table, the phi and Cramer's *V* coefficients yield the same result (0.718), except for the sign. A coefficient of 0.718, whether positive or negative, suggests a strong correlation between gender and mode of transportation. The negative phi coefficient tells us that most of the data fall in the off-diagonal cells. This is easy to see simply by inspecting the contingency table; however, it is the value of the statistic and the *p*-value that indicate whether an association is statistically significant. The range of the phi statistic is ±1. One should ignore the sign when phi is used with nominal data. Cramer's *V* ranges from 0 to 1. The bar chart in Figure 3.20 helps clarify the distribution of the data.

3.4 Kendall's Tau$_b$ and Tau$_c$

The Kendall tau coefficients (τ_a, τ_b, and τ_c) are measures of association between two variables where the scores or values are ranked from highest to lowest, most favorable to least favorable, best to worst, etc. Kendall's tau is

Chi-Square Tests						
	Value	df	Two-Sided Asymptotic Sig.	Two-Sided Exact Sig.	One-Sided Exact Sig.	Point Probability
Pearson chi-square	92.821[a]	1	0.000	0.000	0.000	
Continuity correction[b]	89.923	1	0.000			
Likelihood ratio	102.265	1	0.000	0.000	0.000	
Fisher's exact test				0.000	0.000	
Linear-by-linear association	92.305[c]	1	0.000	0.000	0.000	0.000
Number of valid cases	180					

[a] No cells (0.0%) had expected count less than 5; minimum expected count is 34.22.
[b] Computed only for 2 x 2 table.
[c] Standardized statistic is −9.608.

Symmetric Measures				
		Value	Approximate Sig.	Exact Sig.
Nominal by Nominal	Phi	-0.718	0.000	0.000
	Cramer's V	0.718	0.000	0.000
Number of valid cases	180			

FIGURE 3.18
Results of chi-square tests and phi and Cramer's V tests.

			Transportation		Total
			Personal Vehicle	Campus Transit	
Gender	Female	Count	14	66	80
		Expected Count	45.8	34.2	80.0
		% within gender	17.5%	82.5%	100.0%
	Male	Count	89	11	100
		Expected Count	57.2	42.8	100.0
		% within gender	89.0%	11.0%	100.0%
Total		Count	103	77	180
		Expected Count	103.0	77.0	180.0
		% within gender	57.2%	42.8%	100.0%

FIGURE 3.19
Frequency and percent of mode of transportation by gender.

an alternative measure of association to the Spearman rank order correlation coefficient. This section discusses τ_b and τ_c only. Kendall's τ_a is mentioned only to let the reader know that it is one of the forms but it is fading from use since τ_b provides information for τ_a.

The shape of the contingency table suggests whether to use τ_b or τ_c. The τ_b is recommended for square tables, and it adjusts for tied ranks. Square tables

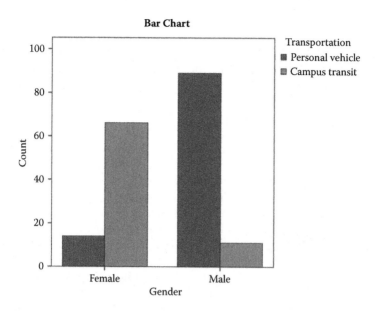

FIGURE 3.20
Distribution of mode of transportation by gender.

have the same number of categories for each variable such as in 2 × 2 and 3 × 3 tables. The τ_c is used for rectangular tables, and it makes no adjustments for tied ranks. The rows and columns of rectangular tables have different numbers of categories, for example, 2 × 3 and 3 × 4 tables.

These methods are known also as the Kendall rank correlation coefficients. They compare the ranked values of two variables to determine agreement (relationship) between the ranks; they do not compare cause and effect. The Kendall tau tests are symmetric. This means that it does not matter which variable is placed in rows or colums and which variable is considered independent or dependent. Like some of the other correlation coefficients, values for τ_b and τ_c coefficients range from –1 to +1, with –1 indicating perfect disagreement between the ranks of two variables and +1 indicating perfect agreement between them.

Perfect agreement or disagreement will occur only for square tables. As the coefficient approaches zero, the agreement or association between the two variables becomes almost nonexistent. Both τ_b and τ_c use the median as their point estimate or location to detect agreement; therefore, their *p*-values will be equivalent. All forms of tau return the same values in square tables. When there are more than two categories for at least one of the variables, τ_c is the appropriate procedure. The formula for τ_c is the difference between the number of higher ranking (concordance) and the lower ranking pairs divided by one half of the sample size squared and multiplied by the minimum number of rows or columns (whichever is less) minus 1 divided by the smaller number of rows or columns. The formula may be written as:

$$\frac{n_c - n_d}{\frac{1}{2}n^2\left(\frac{r-1}{r}\right)}$$

The monotonic relationship (random movement of values in the same or opposite direction) between two variables is assessed by examining the concordance and discordance between the ranks of two variables. If the ranks for both variables are in the same direction, whether in increasing or decreasing order, the variables are said to be concordant (showing agreement between ranks). For example, concordance between the ranks of the measures occurs when a person is ranked higher than another person on one measure and also ranked higher than another person on a second measure. Conversely, discordance between the ranks of the measures appears when the ranks for the two variables are in the opposite direction. A person is ranked higher than another person on one variable and ranked lower on a second variable. In other words, all possible comparisons are made for each case. The null hypothesis for τ_b and τ_c is stated as follows:

H$_0$: The two variables are independent. (Variable names should be given in the statement of hypothesis, e.g., gender and choice of career are independent.)

The alternative hypothesis is stated as:

H$_A$: The two variables are dependent.

Generally, one would not derive the tau statistics by using a hand calculator. However, the following explanation and formulas are provided for those who are interested in the data manipulations. The formula to calculate τ_b uses the difference between the number of concordant and discordant pairs divided by the square root of the sum of the total number of pairs plus the number of tied pairs for one variable. The sum is then multiplied by the sum of the total number of pairs plus the number of tied pairs for the other variable. The formulas for τ_b and τ_c are as follows:

$$\tau_b = \frac{N_c - N_d}{\sqrt{N_c + N_d + \text{Ties}_{\text{var X}}}\,(N_c + N_d + \text{Ties}_{\text{var Y}})}$$

$$\tau_c = \frac{N_c - N_d}{\frac{1}{2}N^2\left(\frac{r-1}{r}\right)}$$

A general formula for tau that is calculated easily is

$$\tau = \frac{2Q}{\frac{1}{2}n(n-1)} - 1$$

where n is the number of pairs of scores and Q is the sum of ranks higher than the preceding ranked case. With a few cases and with one of the variables ranked in increasing order as in Table 3.4, we can use the preceding formula to calculate tau. Example 3.4 illustrates the tau statistic.

Example 3.4

A job analyst wants to know whether a person's job satisfaction and self concept are related. The analyst hypothesizes that the two variables are related, collects data from seven individuals, and ranks their scores on job satisfaction and self concept scales. The analyst then records the scores in a table with one of the variables (job satisfaction) ranked from one to seven. The null hypothesis is stated thus:

H_0: Job satisfaction and self concept are independent.

The alternative hypothesis is stated as follows:

H_A: Job satisfaction and self concept are not independent.

First, calculate the value for Q. Start with the first rank (number 2) for self concept for Person A. Count the number of ranks higher than the number 2 (there are 5). Next count the number of ranks higher than 3 for Person B (there are 4). Count the number of ranks higher than 5 for Person C (there are 2). Continue counting ranks to the right that are higher than each preceding rank. Counting the ranks can be intensive, especially with large samples and/or tied ranks because adjustments must be made for tied ranks. Finally, we sum the number of higher ranks to calculate the concordance between the two variables:

$$Q = 5 + 4 + 2 + 1 + 0 + 1 + 0 = 13; n = 7.$$

TABLE 3.4

Ranks for Self Concept and Job Satisfaction

	A	B	C	D	E	F	G
Job satisfaction	1	2	3	4	5	6	7
Self concept	2	3	5	6	7	1	4

Applying the formula, we have:

$$\tau = \frac{2 \times 13}{\frac{1}{2}7(7-1)} - 1$$

$$= \frac{26}{21} - 1 = 1.238 - 1 = 0.238$$

A tau correlation coefficient of 0.238 shows a positive, but weak, relationship between job satisfaction and self concept. Nevertheless, the analyst is interested in whether the correlation is statistically significant.

Let us use SPSS to calculate tau and the statistical significance for these data. SPSS does not require that the data be entered in rank order; nor is it necessary to be concerned with tied ranks. If the data are on a continuous or ratio scale, SPSS automatically ranks them. The data in Table 3.4 on self concept and job satisfaction will be used to illustrate the following SPSS procedure for producing τ_b and τ_c.

Click Analyze – Descriptive Statistics – Crosstabs – Move job satisfaction to the Row(s) box – Move self concept to the Column(s) box – Click Statistics – Click tau$_b$ and tau$_c$ – Continue – Click Exact – Click Exact within the Exact Tests dialog box – Click Continue – Click OK.

The 0.238 results for Kendall's τ_b and τ_c indicate a very weak relationship. One would not expect such a small correlation coefficient to be statistically significant. The approximate p-value (0.460) and the exact p-value (0.562) are not statistically significant. The analyst must retain the original null hypothesis and conclude that job satisfaction and self concept are independent. The results for τ_b and τ_c are displayed in Figure 3.21.

The previous example of τ_b and τ_c was for a square table depicting self concept and job satisfaction, each with two categories. The τ_c is recommended for irregular tables. Its SPSS path is the same as the path for τ_b, so it will not be repeated here. The data for irregular tables can be entered into the

Symmetric Measures		Value	Asymptotic Standard Error[a]	Approximate T[b]	Approximate Sig.	Exact Sig.
Ordinal by Ordinal	Kendall's τ_b	0.238	0.322	0.740	0.460	0.562
	Kendall's τ_c	0.238	0.322	0.740	0.460	0.562
Number of valid cases		7				
[a] Not assuming null hypothesis.						
[b] Using asymptotic standard error assuming null hypothesis.						

FIGURE 3.21
Results for Kendall's τ_b and τ_c for self concept and job satisfaction.

SPSS spreadsheet as total counts, or the Weight Cases function as shown in Figure 3.4 may be used.

Note that τ_b and τ_c are based on ranked values to produce a coefficient of the association between two variables with two or more levels each. If interest is in observer ratings rather than ranks, the kappa (κ) statistic discussed in Section 3.5 is used.

3.5 Kappa Statistic

The kappa (κ) coefficient measures the extent of agreement between the ratings of two judges or observers on an individual attribute, event, or object. Kappa is an important tool in determining inter-rater reliability. The statistic is based on the difference between two proportions: the observed proportion of consistent ratings by the observers and the expected proportion of ratings by chance. The difference is divided by 1 minus the expected proportion.

Expected proportions are easily calculated when the data are organized in a table. The proportion with observer agreement is the sum of consistent classifications between the two observers (sum of frequencies in diagonal cells) divided by the sample size. The expected proportion is calculated by summing the product of the row 1 total and column 1 total with the product of the row 2 total and column 2 total; then dividing by the sample size squared. After finding the agreement proportion and the expected proportion, the formula is written as: $\kappa = (P_0 - P_e)/(1 - P_e)$. As you can see, κ takes chance into account by using the expected proportion in the calculation.

The κ coefficient generally ranges between 0 and 1. The closer the coefficient is to 1, the higher the level of agreement between the observers. A coefficient near 0 indicates a chance agreement of the observers. A negative κ coefficient means that the agreement between the observers is less than that expected by chance alone.

Essentially, interpretation of the κ coefficient is similar to that of the Pearson product–moment correlation coefficient. Certain general guidelines apply to the interpretation of the coefficient. A coefficient of ≥ 0.80 indicates very high to almost unanimous agreement. A coefficient range from 0.60 to 0.79 indicates high agreement. A range from 0.40 to 0.59 suggests moderate agreement. A coefficient ranging from 0.20 to 0.39 implies fair agreement, and a coefficient ≤ 0.20 indicates very low and almost negligible agreement. Use of the kappa statistic is illustrated in Example 3.5.

Example 3.5

Two administrators at a large urban elementary school classify 88 teachers on whether the teachers are perceived to be student-centered.

TABLE 3.5

Classification of Teachers by School
Administrators

| Administrator 2 | Administrator 1 | | |
	Yes	No	Total
Yes	A	B	
	65	7	72
No	C	D	
	4	12	16
Total	69	19	88

Yes responses indicate student-centered; no responses indicate not-student-centered. Table 3.5 shows the classifications for each administrator. The classifications of Administrator 1 are on the columns, and those of Administrator 2 are on the rows. Notice that the cells are labeled A through D for convenience of calculation only. We will calculate the κ statistic using the data in Table 3.5.

$$\text{Proportion expected: } \frac{72 \times 69 + 16 \times 19}{88^2} = \frac{5272}{7744} = 0.68 \,.$$

$$\text{Proportion with consistent agreement: } \frac{65 + 12}{88} = 0.88 \,.$$

$$\kappa = \frac{0.88 - 0.68}{1 - 0.68} = \frac{0.20}{0.32} = 0.62 \,.$$

The proportion of ratings with consistent agreement appears to be large; however, without a statistical test to show the probability of observed proportion of agreement compared to the expected proportion of agreement, one cannot claim that the proportions are different.

The SPSS procedure for calculating the κ statistic follows the same path as the previous measures of association using the Crosstabs window. Data from the table can be entered using the weight cases method. First, weight the cases; then use the following SPSS path to access the κ statistic.

Click Analyze – Descriptive Statistics – Crosstabs – Move admin2 to the Row(s) box – Move admin1 to the Column(s) box – Click Exact – Click Exact within the Exact Tests dialog box – Click Statistics – Click Kappa – Click Cells (Observed should be selected. Optionally, you can select Expected and Percentages.) – Click Continue – Click OK

The Crosstabulation table shows the classifications by each administrator as recorded in Table 3.5. Examining the Crosstabulation table is a good way to check that the data are entered correctly. Results show a statistically significant κ coefficient ($\kappa = 0.61$, $p < 0.01$). The κ coefficient is consistent with the proportion of agreement ratings (0.88) by the two administrators. Based on the κ coefficient and the p-value, we can conclude that the administrators' ratings of teachers are in strong agreement. Figure 3.22 shows results for the κ coefficient with the Crosstabulation table.

The κ statistic may also be used to assess the ratings of two observers on more than two categories of a variable. For example, if the two administrators in the previous example wished to rate a random sample of 30 teachers on three categories of teacher centeredness (student, parent, administrator), data could be summarized in a table similar to Table 3.6 and then transferred to the SPSS spreadsheet using the Weight Cases method.

The κ statistic is appropriate for showing levels of agreement among judges or observers, not for showing relationships between variables. The Spearman rank-order correlation coefficient (Spearman rho) is used to assess relationships between variables; it is discussed in Section 3.6.

Crosstabulation					
				Administrator	
			1	2	Total
Administrator	1	Count	65	7	72
		Expected Count	56.5	15.5	72.0
		% of Total	73.9%	8.0%	81.8%
	2	Count	4	12	16
		Expected Count	12.5	3.5	16.0
		% of Total	4.5%	13.6%	18.2%
Total		Count	69	19	88
		Expected Count	69.0	19.0	88.0
		% of Total	78.4%	21.6%	100.0%

Symmetric Measures						
		Value	Asymptotic Standard Error[a]	Approximate T[b]	Approximate Sig.	Exact Sig.
Measure of agreement	κ	0.608	0.107	5.740	0.000	0.000
Number of valid cases		88				
[a] Not assuming null hypothesis.						
[b] Using asymptotic standard error assuming null hypothesis.						

FIGURE 3.22
Crosstabulation and kappa coefficient tables for administrator ratings of teachers.

TABLE 3.6

Classification of Teachers into Three Groups by
School Administrators

	Administrator 1		
Administrator 2	Student	Parent	Administrator
Student	xx	xx	xx
Parent	xx	xx	xx
Administrator	xx	xx	xx

3.6 Spearman Rank-Order Correlation Coefficient

The Spearman rank-order correlation coefficient known as Spearman rho (ρ)
is the nonparametric alternative to the Pearson product–moment correlation
coefficient. Like the Pearson technique, the Spearman ρ shows the relationship
of two variables. The Pearson correlation is used for continuous or ratio data
and must meet certain parametric assumptions. The Spearman ρ assumes
only that participants were randomly selected, scores are independent of one
another, and the relationship of one score with another is monotonic.

The Spearman ρ makes no assumptions about the shape of the distribution; consequently it is also appropriate for interval data that do not meet the
normality assumption of the Pearson correlation coefficient. When the data
are on an interval or ratio scale, the procedure ranks the measures for each
variable and then calculates the coefficient on the ranked data.

The Spearman rank-order correlation coefficient is calculated on ranked
data for pairs of scores. Data on two different variables are ranked for each
subject, and the difference between the two ranks is calculated. Each difference is squared, and the squared differences are summed. The squared differences are multiplied by a constant 6, and divided by the number of pairs
multiplied by the number of pairs minus 1. The result is subtracted from 1.
Tied ranks are assigned the average of the two ranks that the scores would
have received if not tied.

The Spearman ρ statistic can be found easily with a hand calculator,
although the procedure becomes more intensive with more than eight or ten
matched pairs. In addition, the coefficient alone does not indicate statistical
significance. If a computer program is not used to calculate the coefficient,
one must consult a table of critical values for the Spearman ρ coefficient to
evaluate statistical significance. For purposes of illustration, the formula to
calculate the Spearman ρ when there are no tied ranks is shown in Figure 3.23.

The values for the Spearman ρ statistic range from –1 to +1. The direction
and values of the statistic are interpreted the same as the Pearson r correlation coefficient. Results near ±1 approach a perfect relationship. If the sign
of the coefficient is positive, then both variables are increasing at the same

$$\rho = 1 - \frac{6\Sigma d^2}{N(N^2 - 1)}$$

FIGURE 3.23
Formula to compute Spearman rank order correlation coefficient where 6 is a constant, Σd^2 is the sum of the squared differences between a pair of ranks for each subject, and N is the number of paired ranks.

time; if the sign of the coefficient is negative, as one variable increases, the other decreases.

Interpretation of the value of the coefficient is somewhat subjective, and one should exercise caution in interpreting the size of the coefficient to mean statistical significance. Statistical significance indicates retention or rejection of the null hypothesis, whereas the size of the coefficient indicates the strength of the correlation. The closer the coefficient is to ±1, the stronger the relationship. One rule for interpreting the strength of a correlation is that coefficients ≥ 0.80 indicate a very strong association; coefficients ≤ 0.30 indicate weak association; and those between 0.30 and 0.80 indicate a moderate association of two variables.

Interpretation of the Spearman ρ statistic leaves ample room for judgment; therefore, one should remember that statistical significance may not indicate the importance of a relationship. In other words, some important relationships can be revealed by nonsignificant correlation coefficients even though the null hypothesis is retained. The null hypothesis for the Spearman ρ procedure is stated as follows:

H₀: No statistically significant association exists between the two variables. (Include the variable names.)

The alternative hypothesis is stated as follows:

Hₐ: A statistically significant association exists between the two variables. (Include the variable names.)

Example 3.6 demonstrates the Spearman ρ procedure.

Example 3.6

A teacher is interested in whether a correlation exists between scores on a mathematics test and scores on an English test for 33 middle school students because she hypothesizes that students who have higher reading and comprehension abilities can read and comprehend mathematics formulas and solve problems accurately. The SPSS path to conduct the Spearman rank-order correlation is as follows.

Click Analyze – Click Correlate – Click Bivariate – Move English and math to the Variables box – Unclick Pearson – Place a check mark in

Correlations			English	Math
Spearman's ρ	English	Correlation coefficient	1.000	−0.168
		Two-tailed sig.	−	0.550
		Number	15	15
	Math	Correlation coefficient	−0.168	1.000
		Two-tailed sig.	0.550	−
		Number	15	15

FIGURE 3.24
Results for Spearman rank order correlation coefficient.

the box next to Spearman – Check that Two-tailed is selected in the Test of Significance section – Check that Flag significant correlations is selected in the lower left of the Bivariate Correlation box – Click OK

Results of the analysis show a nonsignificant association ($\rho = -0.17$, $p = 0.55$) between the scores for English and mathematics. The results for this study are displayed in Figure 3.24.

Chapter Summary

The chi-square test compares the observed and expected frequencies of responses within each category of possible responses for two independent samples. For manual calculations, the chi-square statistic is compared to a table of chi-square values at the appropriate degree-of-freedom and alpha levels to assess the probability of obtaining a value as extreme as the calculated value. Degrees of freedom for the chi-square test of independence are $(r - 1)(c - 1)$, where r is the number of rows and c is the number of columns. If the chi-square value exceeds the table value, then statistically significant differences between the two groups may be claimed. That is, the null hypothesis can be rejected on the basis that responses are not dependent on group membership.

A major difference between Fisher's exact test and the chi-square test is that Fisher's produces a 2 × 2 table (2 rows and 2 columns), and a chi-square test for two independent samples accommodates more than two responses (categories) for a variable. Thus, it is possible to have more than two rows of responses in a table. Expected frequencies for the chi-square test can be calculated by multiplying the total observed responses on one row of interest by the total observed on a column of interest and dividing by the total number of observations.

The phi contingency coefficient is appropriate for 2 × 2 contingency tables. The contingency coefficient is appropriate for larger square tables such as 4 × 4. Cramer's *V* should be used for tables with uneven rows and columns such as a 2 × 3 table. Kendall's tau$_b$ and tau$_c$ are measures of association based on the ranks of independent pairs of scores. Like the Kendall tau statistics, the Spearman rank-order correlation coefficient uses ranked data to assess the association of two variables. Spearman ρ is an appropriate alternative to the Pearson product–moment correlation when bivariate normality cannot be assumed.

Student Exercises

Data files for these exercises can be retrieved online from the file Chapter3data.zip under the Downloads tab at http://www.crcpress.com/ product/isbn/9781466507609. Read each of the following scenarios. Use SPSS to perform the appropriate analyses and respond to the questions.

3.1 A personnel manager is interested in whether job satisfaction is independent of income. For this study, the manager assigned two categories of job satisfaction (high and low) and three categories of income (high, medium, low). Forty-five employees completed job satisfaction questionnaires. The individuals were placed in one of the three income groups (high, medium, or low).

    a. Which one of the procedures of association or agreement is appropriate for this problem? Why did you select the procedure that you selected?

    b. What are the results of the analysis? Give the statistic and *p*-values.

    c. If follow-up tests are necessary, perform the tests and interpret the results.

    d. What might one conclude from this analysis?

3.2 A study was conducted at a middle school to assess the relationship between teachers' use of technology in the classroom and number of student absences. Twenty-five teachers were asked to indicate the number of times they used technology in their classroom and the total number of students who were absent during a five-week period. Use the Spearman rank-order correlation procedure for this problem and respond to the following items.

    a. Write the null hypothesis for this problem.

    b. Report the statistical results of the analysis.

    c. Create a graph of the results.

    d. Write a conclusion statement for the problem including your decision to retain or reject the null hypothesis.

3.3 Two art teachers were randomly selected as judges for an art competition at a local high school. Forty-three students submitted entries. Judging was based on criteria such as originality, creativity, clarity of theme, and overall eye appeal. Judges were asked to classify each art entry as eligible for the district contest using yes or no to indicate the eligibility of a piece to move to the district level. Judges classified each entry independently, and neither judge was permitted to see any classifications of the other judge. Hint: This problem involves determination of inter-rater agreement. In the raw data set, yes = 1 and no = 0. The weight cases data set is also available.

    a. Write the null hypothesis for this problem.

    b. Open the data set and construct a 2×2 table of the judges' responses.

    c. What is the appropriate measure for agreement of the judges?

    d. Perform the analysis and report your results.

3.4 A psychologist hypothesizes that children's interactions with animals are associated with their interactions with people. To test his hypothesis, he engages two independent observers to rank 11 students on their interactions with other children and with animals. During 20-minute observation periods, the children are observed while interacting with each other. They also spend 20-minute periods interacting with puppies and kittens. To avoid the halo effect, one judge ranked each child's interactions with other children, and the other judge ranked each child's interactions with animals. Use the Spearman rank order correlation procedure.

    a. Write the null hypothesis for this problem.

    b. Conduct the analysis and report the value of ρ and the p-value.

    c. Interpret the results of the analysis and write a concluding statement.

3.5 A real estate agent believes that a person's request for a pet-friendly apartment is independent of gender. However, it does appear that more men than women request pet-friendly apartments, so the agent records the number of applicants who apply for an apartment over a three-month period. The agent also collects data from 47 applicants and sorts the data by gender and by requests for pet-friendly apartments. Use the chi-square test to assess this problem of independence.

    a. Write the null hypothesis for this problem.

    b. Open the data set and record the numbers for the independent variables (gender and pet-friendly) in a table.

    c. What are the results of the chi-square test? What are the statistic and p-values?

    d. Write a conclusion for this problem supported by the evidence and include a bar chart.

3.6 Design your own research study to test association or agreement.

    a. Write three or four paragraphs that provide background information for a problem related to your area of study. Use documented sources to support the information.

    b. Write the research question(s) you want to answer.

    c. Write the null hypothesis(es) to be tested.

    d. Identify the appropriate statistic to test your hypothesis(es).

    e. Write a hypothetical results and concluding statement for your problem.

References

Agresti, A. (1992). A survey of exact inference for contingency tables. *Statistical Science*, 7, 131–177. http://www.jst or.org/stable/2246001

Ary, D., Jacobs, L. C., Razavieh, A., and Sorensen, C. K. (2009). *Introduction to Research in Education*, 8th ed. Belmont, CA: Wadsworth Publishing.

Black, T. R. (2009). *Doing Quantitative Research in the Social Sciences: An Integrated Approach to Research Design, Measurement, and Statistics*. Thousand Oaks, CA: Sage.

Costner, H. L. (1965). Criteria for measures of association. *American Sociological Review*, 30, 341–353. http://www.jstor.or g/stable/2090715

DeGroot, M. H. and Schervish, M. J. (2002). *Probability and Statistics*, 3rd ed. Boston: Addison-Wesley.

Gibbons, J. D. and Chakraborti, S. (2010). *Nonparametric Statistical Inference*. Boca Raton, FL: Chapman and Hall/CRC.

Hollander, M. and Sethuraman, J. (1978). Testing for agreement between two groups of judges. *Biometrika*, 65, 403–410. http://www.jstor.o rg/stable/2335221

Knapp, T. R. (1962). Classroom use of rank-order data. *American Statistician*, 16, 25–26. http://www.jstor.org/stable/2681434

Kvam, P. H. and Vidakovic, B. (2007). *Nonparametric Statistics with Applications to Science and Engineering*. Hoboken, NJ: John Wiley & Sons.

Mirkin, B. (2001). Eleven ways to look at the chi-squared coefficient for contingency tables. *American Statistician*, 55, 111–120. http://www.jstor.org/s table 2685997

Sprinthall, R. C. (2012). *Basic Statistical Analysis*. Boston: Pearson.

Stuart, A. (1963). Calculation of Spearman's rho for ordered two-way classifications. *American Statistician*, 17, 23–24. http://www.jstor.org/stable/2682594

Yates, F. (1984). Tests of significance for 2×2 contingency tables. *Royal Statistics Society*, 147 (Part 3), 426–463.

4

Analyses for Two Independent Samples

In Chapter 3, measures of association and agreement between and among two or more variables were discussed. The objective of such procedures is to ascertain the extent to which the variables are related. For some research problems, the objective is to test whether differences exist between unrelated groups.

The objective of testing two independent samples is to determine the extent or degree to which the unrelated groups differ on some variable. For example, one of several procedures for analysis of independent samples may be used for research problems that aim to differentiate scores in reading achievement of boys and girls. Independence of groups and subjects within groups is a research design matter. It is a responsibility of the researcher to assure that observations of one subject are independent of the observations of any other subject in either group. Specifically, one should not be able to tell anything about the scores of one group or subject by knowing the scores of the other group or any other subject.

Selecting independent samples should be carried out in a methodical manner to assure that the samples are selected randomly. Participants can be selected from two different groups or treatments. For example, random samples of boys and girls can be selected from a school system and each group (boys and girls) can be assigned to one of two different teaching methods. A medical research study could select two different samples for research, each sample from a different treatment group.

Nonparametric tests for two independent samples are the analogues for the parametric t-test for independent samples. While the t-test requires continuous data, nonparametric tests for two independent samples are appropriate when the data are collected on rank or nominal scales. In addition, nonparametric procedures for two independent samples are the statistics of choice when data are measured on an interval scale and the assumptions of normality and equal variances cannot be met as required by parametric tests.

This chapter covers six procedures for analyzing two independent samples. The first is the Fisher's exact test to examine dichotomous data for 2×2 tables. Next, the median test that examines equality of two population medians and the Wilcoxon Mann-Whitney U test that assesses differences in ranks of two independent populations are discussed. The chapter also covers the Kolmogorov-Smirnov two-sample test, the Hodges-Lehmann estimate for confidence intervals, and the Moses extreme reaction test.

Formulas and procedures are presented for some of the tests to help readers conceptualize test calculations. The SPSS procedures are included to show steps in the conduct of the analysis for each test, and each test is illustrated with an example. A summary is presented at the end of the chapter. This chapter concludes with student exercises to reinforce learning and practice of the tests presented.

The data sets for the examples and exercises of this chapter can be retrieved online from the file Chapter4data.zip under the Downloads tab at http://www.crcpress.com/product/isbn/9781466507609.

4.1 Fisher's Exact Test for 2 × 2 Tables

The purpose of Fisher's exact test for 2 × 2 tables is to evaluate differences between two discrete dichotomous variables, where responses of two independent groups fall exclusively into one category or the other. One of two possible responses (usually coded as a 0 or a 1 for statistical analysis) can be recorded in a 2 × 2 table for every participant in each group. The 2 × 2 table is called a contingency table. The number in each of the four cells represents the frequency of occurrence or frequency of observing a specific response. The objective of the test is to ascertain whether two groups differ in the proportions in which the responses are observed.

This procedure is useful especially for small sample sizes ($N \leq 20$) when the data are measured on ordinal or nominal scales. Since each participant in each group can be classified on only one of two responses (yes or no, pass or fail, etc.), it is possible that some participants may not be classified in one of the cells. Thus, 2 × 2 tables for Fisher's exact test can have zero frequencies in at least two cells. Consequently, cells can contain unequal N values.

The Fisher exact test for 2 × 2 tables produces the exact probability of obtaining a set of frequencies as extreme or more extreme than those observed. It is assumed that the margin totals on the 2 × 2 table are fixed, although this is not a strict requirement of the test. Suffice it to say that since the margins are fixed, after a frequency is recorded in one cell for a row or column, frequencies for the other cell on the row or column are predetermined.

Calculations are based on the hypergeometric distribution—a discrete distribution that produces the probability of observing a specific number of frequencies in a given number of trials without replacement (each observation may be counted only once). A response classified into one cell (group or category) cannot be classified into any other group or category.

The formula for calculating the probability of observing such frequencies can require manipulation of extremely large numbers (even with small sample sizes), making hand calculations complex and time consuming. If none of the cells is zero, the analysis will need to be conducted an appropriate number of times to account for the probability of extreme occurrences. The probabilities from all analyses are summed to calculate the probability level that will indicate whether the null hypothesis can be rejected. Conceptually, the test sums the probabilities of all cells with frequencies equal to or larger than the value entered into the first cell. The formula in Figure 4.1 for Fisher's exact test shows the sums of the cells on the rows and columns.

The null hypothesis tests the probability of the occurrence of frequencies more extreme than those observed. Example 4.1 will help illustrate the cumbersome nature of calculations for Fisher's exact test.

Example 4.1

A market researcher is interested in assessing whether eight individuals (four females and four males) differ in their car selections (foreign or domestic) when given a choice. The first step is to construct a 2 × 2 contingency table as shown in Table 4.1, and record the appropriate number in each cell. The entries A, B, C, and D are shown on the table for use of the formula in hand calculations only; otherwise, the cells would not be labeled.

Next, apply the formula in Figure 4.1. Results of the calculations indicate that the p-value is 0.428; however, the calculations are not completed. It is possible to have a 0 in one of the cells. Notice that there is already a 1 in one of the cells. Using the same example and without changing the margin totals, it is obvious that even more extreme values could occur. To test the probability of observing random case assignments more extreme than those in Table 4.1, the sum of the probability of observing the case assignments in Table 4.1 and the probability of observing the more extreme occurrence of random assignment shown in Table 4.2 must be calculated.

The probability of observing cases in Table 4.2, which are more extreme than those in Table 4.1, is calculated in the same way as the first calculations. For example, instead of the distribution of responses shown in Table 4.1, we could observe the frequency of cases shown in Table 4.2.

$$p = \frac{(A+B)!(C+D)!(A+C)!(B+D)!}{N!A!B!C!D!}$$

FIGURE 4.1
Formula to calculate Fisher's exact test.

TABLE 4.1

Data for Fisher's Exact Test Calculations for Car Preference by Gender

	Foreign Car	Domestic Car	Total
Male	(A) 3	(B) 1	4
Female	(C) 2	(D) 2	4
Total	(A + C) 5	(B + D) 3	8

TABLE 4.2

Data for Fisher's Exact Test Calculations for
Car Preference by Gender: Extreme Case

	Foreign Car	Domestic Car	Total
Male	(A) 1	(B) 3	4
Female	(C) 4	(D) 0	4
Total	(A + C) 5	(B + D) 3	8

The resulting p-value for data in Table 4.2 is 0.0714. Next the two probabilities need to be summed to determine the probability of observing cell
counts as extreme as or more extreme than those observed in Table 4.1.
As we can see, the lower right cell contains a 0, indicating that no females
preferred domestic cars. Thus, the probability of observing cases as they
fall in Table 4.1 or observing more extreme responses would be the sum
of the probabilities for Tables 4.1 and 4.2:

$$p = 0.4286 + p = 0.0714 = 0.50.$$

These results tell us that males and females do not differ in their choices
of cars when they can select a foreign or domestic car.

At this point, the laborious nature of using hand calculations to determine
the probability of observing more extreme values than those in any given
table should be obvious. Specifically, if the lowest value in a table is 2, it is
logical that a more extreme value of 1 could be observed. If the lowest value
in a table is 3, we could observe the same margin totals with an extreme value
of at least 2 in one of the cells; an extreme value of 1 in at least one of the cells;
or an extreme value of 0 in one of the cells. All these scenarios are possible
when the most extreme values are missing from a contingency table. In these
cases, separate exact probability levels must be calculated for each possible
extreme value within its own contingency table, and the separate exact probability levels should be summed to derive the probability of observing a
given contingency table or one with more extreme values. Thus, for a table
with the lowest value of 2, one would need to calculate 3 exact probabilities.
If the lowest value in a table is 3, one would need to calculate 4 exact probabilities—one for the lowest value of 3; one for the lowest value of 2; one for
the lowest value of 1; and one for the lowest value of 0. These calculations can
be extremely cumbersome and tedious using a hand calculator! Example 4.2
is another illustration of Fisher's exact test.

Example 4.2

Data were collected by medical researchers at a regional clinic to investigate whether more men or more women patients will volunteer to
participate in an experimental study for clinical trial research on sleep

TABLE 4.3

Data for Agreement to Participate in
Experimental Study by Gender.

	Female	Male
Agree	8	2
Not agree	4	9

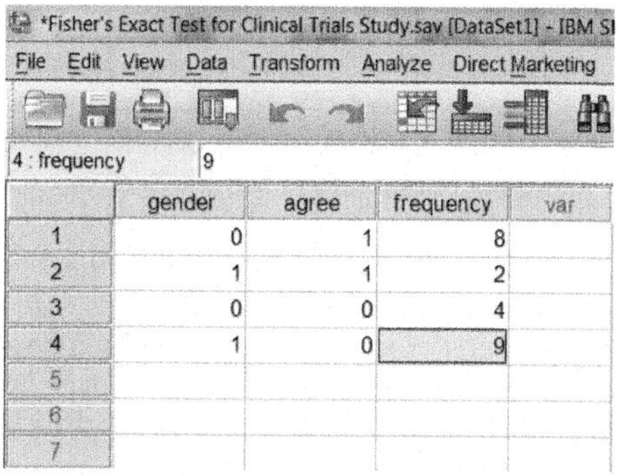

FIGURE 4.2
SPSS Data view for weighted cases.

disorders. The researchers selected 11 men and 12 women patients for
the study. Patients were asked whether they were willing to take part
in an experimental clinical trial study on sleep disorders. Each patient
responded yes for agree to participate in the study or no for not agree
to participate in the study. The researchers were interested in the
probability of observing obtained frequencies or values more extreme
than those in Table 4.3.

The null hypothesis for this study tests the exact probability that the
number of cases within each cell is random. The null hypothesis can also
be stated as follows:

H_0: Patient participation in clinical trials is independent of gender.

For this study, the researchers are interested in the one-sided p-value.

Figure 4.2 displays the data in the SPSS spreadsheet using the Weight
Cases method. Gender is coded 0 for women and 1 for men. Agreement
to participate in the study is coded 0 for not agree and 1 for agree. The
probability can be easily and quickly calculated with the SPSS software.

The SPSS path for the Fisher's exact test is the same as the one for
the chi-square test using the Crosstabs procedure in the Descriptive
Statistics drop-down menu. The steps are repeated here for convenience.
The weight cases method is suggested unless you are entering raw data.
Remember to weight the cases before conducting the analysis by clicking the Data tab, selecting the option for Weight cases by and moving
frequency to the Frequency Variable box.

Comparison variables are usually coded as 0 and 1. Group and response variables are set to Nominal Scale in the Variable View of the spreadsheet. See Chapter 3 (Figure 3.3) to review the dialog box for Weight Cases. Other SPSS dialog boxes for Crosstabs are presented in the figures for Chapter 3. These dialog boxes are not repeated in this chapter. Following is the SPSS Path for Fisher's exact test.

Click Analyze – Click Descriptive Statistics – Click Crosstabs – Move gender to row(s) and Move agree to column(s) – Click the Exact button and then click Exact – Click Continue – Click Statistics – Select Chi-square – Click Continue – Click Cells – Observed in the Counts section will be selected already – Select Expected – Select Row and Column in the Percentages section – Click Continue – (Optionally, check Display clustered bar charts to create a bar chart of the 2 × 2 table) – Click OK

Figure 4.3 displays the observed and expected frequencies in each cell. Figure 4.4 reports the results for the Fisher's exact and the chi-square tests. The result of Fisher's exact test is reported by default. One should interpret the p-value (0.03 for the one-sided test) for a 2 × 2 table and a small sample size. The first note (a) tells us that the data do not meet the expected frequency per cell for a chi-square test.

Interpreting either the chi-square p-value (0.03 for the one-sided test) or the Fisher's exact test p-value, we would reach the same conclusion for this example. However, this may not always be the case, especially for small sample sizes and data that are not normally distributed. Note that Fisher's exact test produces only probability levels, not test statistics. The probability levels are 0.03 for the one-sided test and 0.04 for the two-sided test for this example.

			Not Agree to Experiment	Agree to Experiment	Total
Gender	Female	Count	4	8	12
		Expected count	6.8	5.2	12.0
		% within gender	33.3%	66.7%	100.0%
		% within agree to experiment	30.8%	80.0%	52.2%
	Male	Count	9	2	11
		Expected count	6.2	4.8	11.0
		% within gender	81.8%	18.2%	100.0%
		% within agree to experiment	69.2%	20.0%	47.8%
Total		Count	13	10	23
		Expected count	13.0	10.0	23.0
		% within gender	56.5%	43.5%	100.0%
		% within agree to experiment	100.0%	100.0%	100.0%

FIGURE 4.3
Observed and expected frequencies for agreement by gender. Crosstabulation: gender and agreement.

The researchers should reject the null hypothesis that patient agreement to participate in a clinical trial study on sleep disorders is independent of gender. They can conclude that there is a statistically significant association of gender with willingness to participate in the study. More women (8) than men (2) agreed to participate. A bar chart showing the number of men and women who agreed to participate in the clinical trial is displayed in Figure 4.5.

Chi-Square Tests						
	Value	df	Two-Sided Asymptotic Sig.	Two-Sided Exact Sig.	One-Sided Exact Sig.	Point Probability
Pearson chi-square	5.490[a]	1	0.019	0.036	0.026	
Continuity correction[b]	3.694	1	0.055			
Likelihood ratio	5.785	1	0.016	0.036	0.026	
Fisher's exact test				0.036	0.026	
Linear-by-linear association	5.251[c]	1	0.022	0.036	0.026	0.024
Number of valid cases	23					

[a] One cell (25.0%) had expected count less than 5; minimum expected count is 4.78.
[b] Computed only for 2 × 2 table.
[c] Standardized statistic is −2.292.

FIGURE 4.4
Results of Fisher's exact test for agreement to participate in clinical trial study by gender.

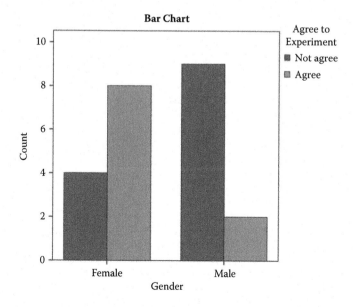

FIGURE 4.5
Bar chart showing agreement to participate in experiment by gender.

4.2 Median Test

The median test, also known as Mood's median test for two independent samples, compares the medians of two different populations. Data are categorized into one of two groups: one group includes values equal to or less than the median of all the data combined; and the other group includes values that are greater than the median of all the data combined. In other words, the position of each value within a set of values is evaluated against the median value for all of the values in the two groups combined.

Data should be on at least an ordinal scale. One advantage of the median test is its insensitivity to departures from homogeneity of variance. No assumptions of data distributions across groups need to be met; however, the test requires that samples be drawn randomly from the populations, and samples should be independent of one another. The median test can be applied to more than one group as discussed in Chapter 5. The focus in this section is use of the median test to compare two groups.

The median test is useful for studies that investigate individual reactions to certain situations, such as medications, diets, therapies, and exercise programs. Although the test is not used as often as some of the other nonparametric tests of differences between two or more independent groups, it is nonetheless a useful tool that warrants discussion.

Example 4.3 illustrates a problem using the median test.

Example 4.3

A counselor for individuals classified as illegal substance abusers is interested in the effectiveness of the type of counseling they receive. All the clients participated in weekly group or individual counseling sessions for the previous three months. The counselor selected a subgroup of 20 clients from a larger pool of clients who completed the counseling sessions within the last month. Ten of the 20 were assigned randomly to the individual counseling group, and the remaining ten were assigned to group counseling sessions. Both groups included only adult males who had been addicted to illegal substances for 6 to 12 months prior to treatment.

The counselor was interested in the number of relapses that a client experienced in the first two weeks after completing the counseling sessions. In individual follow-up meetings, the counselor asked each client to report the number of times he used an illegal substance within two weeks after completing counseling. The counselor hypothesized that there would be no difference in the number of relapses for clients based on type of counseling (group or individual) received.

SPSS offers two options to perform the median test. The following procedure accommodates comparisons of the median values for more than two groups. However, this option can also be used to compare only two groups. The statistical results show a statement of the null hypothesis tested, name of test, p-value, and decision rule. This option is shown in the following SPSS path for two independent samples.

Click Analyze – Click Nonparametric Tests – Click Independent Samples – Click Customize analysis on the Objective tab – Click the Fields tab – Move use to the Test Fields box – Move countype (counseling type) to the Groups box – Click the Settings tab – Click Customize Test – Select Median Test (k samples) in the Compare Medians across Groups panel – Leave the Pooled samples median selected – Select None in the Multiple comparisons box – Click Run

Figure 4.6 displays the SPSS path for the median test for two independent samples.

Figures 4.7 through 4.9 display important SPSS dialog boxes for customized tests. In this case, the median test is selected. Figure 4.7 shows three tabs (Objective, Fields, and Settings) that allow customization of an analysis. Also, Figure 4.7 displays a description in the lower panel of the dialog box that lists the kinds of customized analyses possible. Figure 4.8 shows the Fields tab activated with the appropriate fields selected. Figure 4.9 shows the open Settings tab with the settings selected for the median test. Note that this dialog box allows one to select other customized tests as well.

As shown in Figure 4.10, a researcher can confidently retain the null hypothesis of no difference in number of relapses (illegal substance use) between clients in group counseling and those in individual counseling for the first two weeks after counseling. To report the results of this analysis,

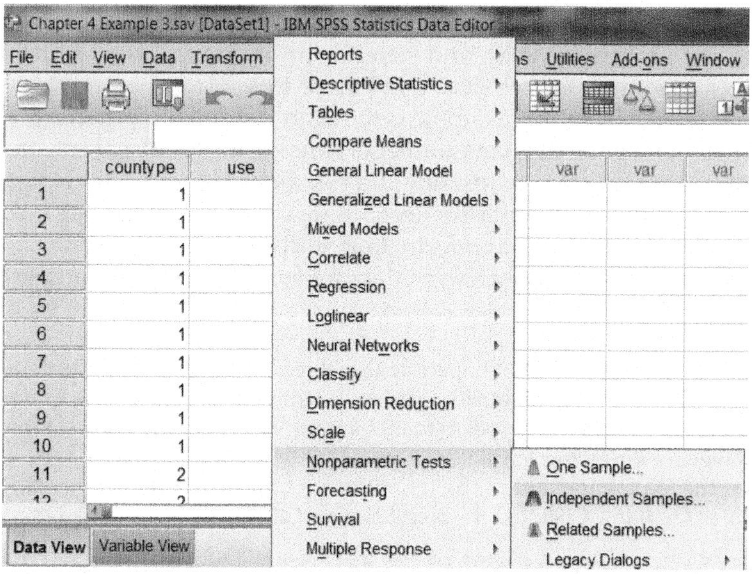

FIGURE 4.6
SPSS path for median test for two independent samples.

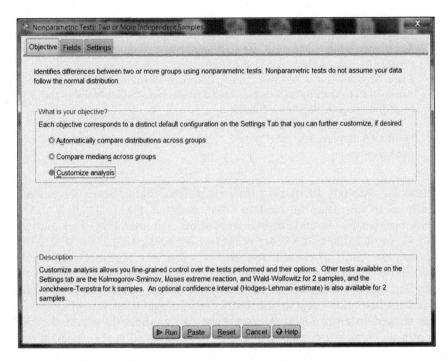

FIGURE 4.7
SPSS dialog box for two or more independent samples with Objective tab open.

the researcher would simply restate the null hypothesis, give the name of the test, the probability level, and state the decision rule. The results of the analysis for Example 4.3 are shown in Figure 4.10.

The decision rule in this example is to retain the null hypothesis ($p = 0.66$) and conclude that the median number of times using an illegal substance in the first two weeks after counseling did not differ between the two groups. A double click on the Hypothesis Test Summary box will generate box plots for the groups and statistical information as displayed in Figure 4.11.

The second SPSS path to analyze data for two independent samples is as follows.

> Click Analyze – Click Nonparametric Tests – Click Legacy Dialogs – Click K Independent Samples – Move use to the Test Variable List – Move countype (counseling type) to the Grouping Variable box – Click Define Range for Grouping Variable – Type 1 for the Minimum value – Type 2 for the Maximum value – Click Continue – Unclick Kruskal-Wallis H – Click Median – Click Exact – Select Exact – Click Continue – Click OK

As one would expect, results of the median test are the same as the results using the first procedure (median = 5.50, exact *p*-value = 0.66). A helpful feature of this procedure is that the results give the number of individuals who

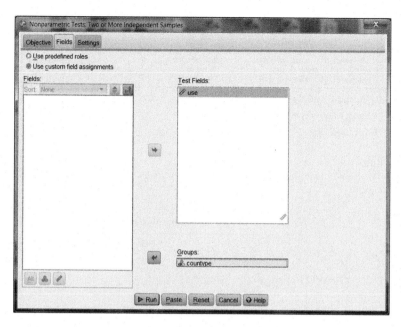

FIGURE 4.8
SPSS dialog box for two or more independent samples with Fields tab open.

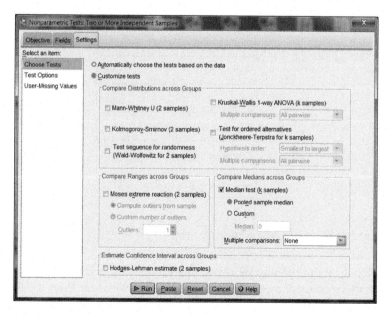

FIGURE 4.9
SPSS dialog box for two or more independent samples with Settings tab open.

Hypothesis Test Summary

	Null Hypothesis	Test	Sig.	Decision
1	The medians of number of times using are the same across categories of counseling type.	Independent-Samples Median Test	0.6563[a,b]	Retain the null hypothesis.

Asymptotic significances are displayed. The significance level is .05.

[a] Exact significance is displayed for this test.

[b] Fisher Exact Sig.

FIGURE 4.10
Results of median test for individuals using illegal substances following two different counseling methods.

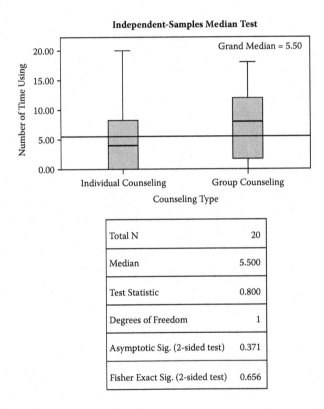

FIGURE 4.11
Box plots and statistical information for median test.

are at or below the median or above the median by group. Four clients in the individual counseling group had relapses (use) greater than the median relapse of 5.5; six clients who received group counseling had use greater than the median. Six clients in the individual counseling group had use equal to or less than the median, and four clients in the group counseling sessions had use equal to or less than the median. Results of the analysis are shown in Figure 4.12.

There is still another piece of information that the researcher may want in an effort to fully understand the data: effect size. In this case, with the large *p*-value, the type of treatment (group or individual counseling) had very little effect on use of illegal substances between the groups following their participation in counseling sessions. In such cases, calculating an effect size is unnecessary. The procedure is presented here for illustration purposes. In addition, a researcher may want to include a chart when presenting results. The following SPSS path will produce the chi-square value, effect size, and a chart showing differences in use by group.

First, two groups must be created and coded by group. A 0 may be used to code the group that has values equal to or less than the median, and a 1 may be used to code the group that has values greater than the median. Use the following SPSS path to form two dichotomous groups.

Click Transform – Click Recode into Different Variables – Move use to Numeric Variable box – Type a name in the Output Variable Name box (reuse is the new variable name typed in the Name box for this example) – Click Change – Click Old and New Values – Select Range Lowest

Frequencies

		Countype	
		Individual counseling	Group counseling
Use	> Median	4	6
	≤ Median	6	4

Test Statistics[a]

	Use
N	20
Median	5.50
Exact Sig.	0.656
Point Probability	0.477

[a] Grouping Variable: countype

FIGURE 4.12

Results of median test showing number of individuals above and below median for two counseling methods.

> through value and type 5.50 in the box (because 5.50 is the median) –
> Select Value in the New Value panel – Type 0 in the box – Click Add –
> Click All other values – Select New Value and type 1 in the New Value
> box – Click Add – Click Continue– Click OK

Figure 4.13 displays the SPSS dialog box for recoding variables. The new
variable (reuse) will appear in the Data View of the SPSS spreadsheet and
use values of 0 or 1 to indicate the two groups. Now the analysis can be conducted to produce the effect size and a chart. Following is the SPSS path for
this procedure.

> Click Analyze – Click Descriptive Statistics – Click Crosstabs – Move the
> grouping variable (countype) to the Rows box – Move the newly created
> variable (reuse) to the Columns box – Click Exact –Select Exact – Click
> Continue – Click Statistics – Select Chi-square – Click Phi and Cramer's
> V – Click Continue – Click Display Clustered Bar Charts in the Crosstabs
> dialog box – Click OK

The researcher should report the same information as in the previous analyses in addition to the effect size (phi and Cramer's $V = 0.20$), which is a small
effect size. Figure 4.14 displays the division of the two groups by the median
in the Crosstabulation table and the results of the phi and Cramer's V statistics. The bar chart in Figure 4.15 shows the number of times above and
below the median that a client used an illegal substance by type of counseling group.

    The median test assesses the difference in the medians of two groups
by location of observations within each group relative to the median of

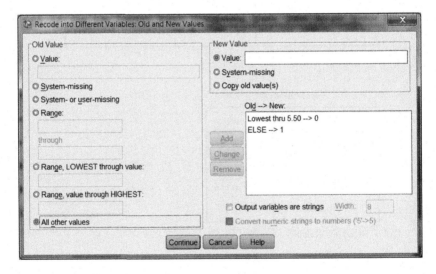

FIGURE 4.13
SPSS dialog box to recode into different variables.

Countype–reuse Crosstabulation

Count

		Reuse		Total
		0	1	
Countype	Individual counseling	6	4	10
	Group counseling	4	6	10
Total		10	10	20

Symmetric Measures

		Value	Approx. Sig.	Exact Sig.
Nominal by Nominal	Phi	0.200	0.371	0.656
	Cramer's V	0.200	0.371	0.656
N of Valid Cases		20		

FIGURE 4.14
Crosstabulation table for number above and below median and phi and Cramer's *V* effect size statistics.

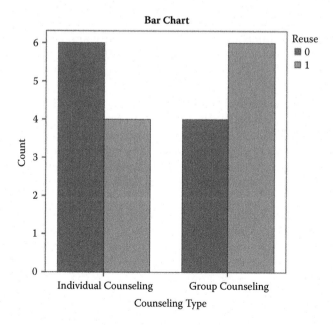

Bar Chart

FIGURE 4.15
Bar chart showing use of illegal substance for groups above and below median by type of counseling.

the combined groups. One group will have observations at or below the median and the second group will have observations above the median. The Wilcoxon-Mann-Whitney U test is considered more powerful than the median test because the rank of each observation is considered instead of only the relation of a score to the median value in the distribution. The Wilcoxon-Mann-Whitney U test is discussed in Section 4.3.

4.3 Wilcoxon-Mann-Whitney U Test

The Wilcoxon-Mann-Whitney U test is a counterpart to the parametric independent samples t-test for evaluating whether two samples were drawn from two different populations with equal means. The test is also known as the Wilcoxon Rank-Sum test, the Mann-Whitney U test, the Mann-Whitney test, or simply the U test. The Mann-Whitney test is especially useful for small data sets. Variables should be measured on at least an ordinal scale. The requirement of interval data is not necessary; however, the test will accommodate interval data (continuous scores) and data measured on an arbitrary scale.

For example, test scores are usually measured on continuous scales, and variables such as beauty, pride, and loyalty are measured on arbitrary scales. If the variables are on a continuous scale, the Mann-Whitney procedure ranks the observations automatically when computer software is used. Other assumptions of parametric tests, such as normality and equal variances, do not need to be met. However, the assumptions of independence of observations and random sample selection should be met to assure a non-biased sample.

The Mann-Whitney test examines the medians of the distributions for two samples to determine whether the median of one sample is larger (or smaller) than the median of the other sample. Scores from both groups are combined and ranked in order of increasing size. The smallest score is ranked 1; the second smallest score is ranked 2, etc. The algebraic sign associated with the score (+ or –) is considered in the ranking so that negative scores are given lower ranks than positive scores.

The ranks in one group, say Group X, are summed and noted as the Wilcoxon value (W_X). Next, the ranks in the other group, say Group Y, are summed. These ranks may be noted as W_Y. It is very important that as the scores are ranked, they retain their original group membership. The sum of the ranks for the two groups should be the same as the sum of the ranks for the combined groups.

The null hypothesis states that the ranks for the two groups are equal, or the null hypothesis may state that there is no difference in the ranks for the two groups. Under a true null, the average ranks for the groups should be approximately equal. However, if the sum of the ranks for one group is either very large or very small when compared to the sum of the ranks for the other group, one may conclude that the samples were taken from two different populations.

The following four steps produce the Mann-Whitney statistic for small sample sizes (< 20). The steps are included for illustration purposes only to help demonstrate how data are ranked and how the statistic is calculated. Researchers most likely will use computer software such as SPSS to make the calculations.

1. Rank the data by increasing values and label each rank by its group name or as an X or a Y so that the original identity of the data is not lost in the process.

2. Sum the ranks for each group.

3. Check that the sums of the ranks for the two groups combined are equal to the sum of the ranks for the combined group. Table 4.4 displays data for the first three steps of the procedure to calculate the Mann-Whitney test statistic. The sum of the ranks for group X is 49 (n = 7). The sum of the ranks for group Y is 71 (n = 8). The sum of the ranks for the two groups is 49 + 71 = 120. Calculate the sum of ranks for both groups using the following formula:

$$\text{Sum of ranks for Group X} + \text{sum of ranks for Group Y}$$
$$= N (N + 1)/2 = 15(15 + 1)/2 = 240/2 = 120.$$

TABLE 4.4

Ranked Data for Two Combined Samples (X and Y)

Group	Score	Rank
X	56	1
X	62	2
X	70	3
Y	71	4
Y	78	5
Y	80	6
X	81	7
Y	83	8
Y	84	9
X	86	10
Y	88	11
X	91	12
Y	92	13
X	95	14
Y	97	15

4. Conduct the statistical test for the null hypothesis. Recall that the null hypothesis is that two groups from different distributions have the same mean. In other words, the null hypothesis tests the probability that the average ranks for Group X are equal to the average ranks for Group Y.

If there is no difference between the average ranks for the two groups, then the average group ranks should be approximately equal. However, if the sum of the ranks is quite different in size, then we may suspect that the two groups were not from the same distribution. The two-tailed test is used if we do not wish to predict the direction of the difference.

For the two-tailed test, the alternative hypothesis states that the probability that Group X is equal to Group Y ≠ ½. The expectation is that the groups will have equal distributions; thus ½ is used in the alternative hypothesis. The test evaluates whether the probability is > ½: that a score from Group X is larger than a score from Group Y. As demonstrated previously, hand calculations may be easy with a few cases, but with more than 10 cases, it is not efficient to depend on manual calculations and statistical tables to interpret results.

An explanation of the algorithm for calculating the value of the Wilcoxon-Mann-Whitney test statistic is based on the following formula:

$$W = n_{small} \times n_{total} - \text{rank sum of } n_{small}$$

where W is the value to locate on a table of critical values at the appropriate sample sizes and alpha level to evaluate statistical significance; n_{small} is the number of cases in the smallest sample; n_{total} is the number of cases in the combined samples; and rank sum of n_{small} is the sum of the ranks for the smallest sample size. In practice, no one would use hand calculations and a table of critical values to determine statistical significance when statistical software will yield the Mann-Whitney U statistic and level of significance. Example 4.4 illustrates the Wilcoxon-Mann-Whitney U test procedure using SPSS.

Example 4.4

A teacher is interested in the number of trials required by students who have been instructed on transfer of knowledge to solve mathematical problems. The experimental group (*n* = 5) participated in a two-hour workshop on transfer of knowledge in problem solving. The control group (*n* = 5) did not participate in the workshop. The teacher's objective was to test whether there was a difference in the number of trials the groups required to solve the problems on a mathematics test.

The research question for this study is stated as follows: To what extent is there a difference in the number of trials to solve a set of mathematics problems for students who participated in a workshop on transfer of knowledge and those who did not participate in the workshop? The null hypothesis is stated as follows:

H_0: No statistically significant difference exists in the number of trials it takes to solve a set of mathematics problems for students who participated in a workshop on transfer of knowledge and those who did not participate in the workshop.

SPSS will calculate the Mann-Whitney U test quickly and accurately. The SPSS path is as follows.

Click Analyze – Click Nonparametric Tests – Click Legacy Dialogs – Click 2 Independent Samples – Move trials to the Test Variable list – Move Group to the Grouping Variable list – Click Define Groups and type a 1 in the Group 1 box and type a 2 in the Group 2 box - Click Continue – Click Exact Tests – Select Exact – Click Continue – Be sure to click the box next to Mann-Whitney test if it is not already selected – Click OK

Figure 4.16 displays the SPSS dialog box for the two-independent samples tests with the Mann-Whitney U test selected.

Results of the analysis revealed a statistically significant difference between the medians of the control and experimental problem-solving groups (U = 3.00, p = 0.030). The control group had the higher mean rank (8.00), indicating that the control group required a higher number of trials than the experimental group (mean rank = 3.60) for solving the mathematics

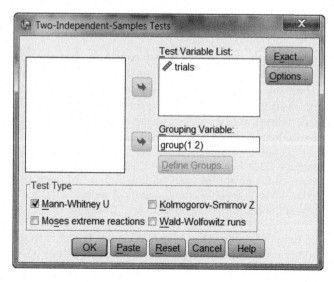

FIGURE 4.16
SPSS dialog box for two independent samples tests with Mann-Whitney U test selected.

Ranks

	Group	N	Mean Rank	Sum of Ranks
	Experimental	5	3.60	18.00
Trials	Control	6	8.00	48.00
	Total	11		

Test Statistics[a]

	Trials
Mann-Whitney U	3.000
Wilcoxon W	18.000
Z	−2.191
Asymp. Sig. (2-tailed)	0.028
Exact Sig. [2 × (1-tailed Sig.)]	0.030[b]
Exact Sig. (2-tailed)	0.030
Exact Sig. (1-tailed)	0.015
Point Probability	0.006

[a] Grouping Variable: group

[b] Not corrected for ties.

FIGURE 4.17
Results of Wilcoxon-Mann-Whitney U test for number of trials to solve mathematics problems by group.

problems. Consequently, the teacher can conclude that the experimental group that received the instruction on transfer of learning performed at a higher level than the control group that received no instruction on transfer of learning. Results of the analysis are shown in Figure 4.17. A bar chart illustrates the distribution of number of trials for each group. Use the following path to create the bar chart in Figure 4.18.

> Click Graphs – Click Legacy Dialogs – Click Error bar – Select Simple and Summaries for groups of cases – Click Define – Move trials to the Variable box – Move group to the Category Axis box – Click Titles to input a title at the top or bottom of the chart – Click OK

An alternative way to perform the Mann-Whitney U test is to use the SPSS path illustrated for the median test. Select Mann-Whitney (2 samples) in the Customize tests panel on the Settings tab. The steps of the procedures are shown in Figure 4.6 and accompanying relevant dialog boxes appear in Figures 4.7 through 4.9.

Results show the null hypothesis tested, the probability level, and the decision or conclusion from the test. A simple frequency chart and statistical results can be obtained by double clicking on the hypothesis test summary output. Figure 4.11 for the Median test displays the kind of information available. Additional information for the Mann-Whitney U test is not shown here. Using both SPSS paths is not a bad idea, especially if one is unsure about the interpretation of the results (whether to retain or reject the null hypothesis). Figure 4.19 shows the results when using this alternative method.

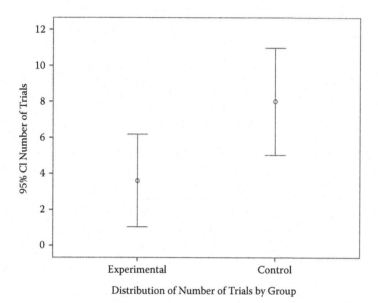

Distribution of Number of Trials by Group

FIGURE 4.18
Error bar chart for distribution of number of trials by group.

Hypothesis Test Summary

	Null Hypothesis	Test	Sig.	Decision
1	The distribution of number of trials is the same across categories of group.	Independent-Samples Mann-Whitney U Test	0.0303[a]	Reject the null hypothesis.

Asymptotic significances are displayed. The significance level is 0.05.

[a] Exact significance is displayed for this test.

FIGURE 4.19
Results of Mann-Whitney U test.

The Wilcoxon-Mann-Whitney U test is a more powerful option than the median test that simply assesses the difference between two medians without regard for the order of the raw data. The Kolmogorov-Smirnov test is another alternative to the parametric independent samples t-test. In addition to evaluating differences between two medians, the Kolmogorov-Smirnov test for two independent samples assesses the similarity of the shapes of the two distributions. Section 4.4 presents information about the test.

4.4 Kolmogorov-Smirnov Two-Sample Test

The Kolmogorov-Smirnov (K-S) test to check for normality of distributions is a common practice for parametric statistical procedures. The K-S test for normality is a modification of the K-S two-sample test for nonparametric statistics. The K-S test for two independent samples evaluates whether two samples were drawn from populations with similar distributions. The K-S test, like the Wilcoxon-Mann-Whitney U test and the median test, provides another alternative to the parametric independent samples t-test. The K-S test assesses differences between two means or medians and provides information about the shape of the distributions.

The purpose of the two-sample K-S procedure is to test the agreement between two samples coming from two different cumulative distributions. The two-tailed test detects differences such as central tendency, variance, and skewness between two samples; the one-tailed test indicates whether the values from one sample are stochastically (randomly) larger than the values of the other sample. For example, the two-tailed test may be used to assess whether scores of an experimental group are higher (or lower) than scores of a control group. The test also assesses whether two samples coming from the same distribution have the same or similar shapes.

The K-S test is based on the idea that two samples drawn from the same population will have similar cumulative distributions. If the cumulative distributions are far apart, we may assume that the samples came from different distributions. In this case, we must reject the null hypothesis that the samples came from the same distribution. The K-S procedure requires the calculation of a cumulative distribution for each sample.

First, the values are ordered for the two samples into two different arrays such that the scores are recorded from lowest to highest in value for each sample. Next, a separate column or row can be used to calculate the cumulative distribution for each sample. The first cumulative score (x_1) is that score value, the second cumulative score (x_2) is the first cumulative (x_1) plus the second score. The third cumulative (x_3) is the second cumulative (x_2) plus the third score, and so on until all the cumulative scores (x's) are recorded for each sample.

TABLE 4.5

K-S Statistic Calculations for Comparison of Two Weight Loss Plans for
Two Independent Groups

Plan 1					
Pounds lost	0.2	0.7	1.2	1.5	2.1
Cumulative pounds lost	0.2	0.9	2.1	3.6	5.7
Percent cumulative pounds lost	0.035	0.159	0.368	0.631	1.00
Plan 2					
Pounds lost	0.4	1.5	2.8	3.4	3.5
Cumulative pounds lost	0.4	1.9	4.7	8.1	11.6
Percent cumulative pounds lost	0.034	0.164	0.405	0.698	1.00
Difference in percent	0.001	0.009	0.037	0.067	0.000

After the cumulative distributions are calculated for each sample, the percent of each cumulative score (x) based on the total cumulative score is recorded for each sample. Finally, the difference between the percents of each of the cumulative scores for the two groups is calculated and the greatest difference in percents is identified. The null hypothesis is based on the probability of obtaining a difference as great as or greater than the observed difference. For a two-tailed test, the maximum absolute difference is found; whereas for a one-tailed test, the maximum difference in the predicted direction is of interest.

Whether a test is one-tailed or two-tailed, the sampling distribution of the differences between the respective samples are known. The algorithm for the two-tailed test uses the absolute value of the largest difference in either direction to show differences between the samples. Hand calculations can be a bit intensive, especially for a large number of cases involving large values. The ten cases displayed in Table 4.5 illustrate calculations for the K-S test and were taken from Example 4.5.

Example 4.5

A dietitian proposes two different weight loss plans. The dietitian's hypothesis is that the population distribution for individuals participating in plan 1 is identical to the population distribution for individuals participating in plan 2. Data are collected from 20 individuals who volunteered to be assigned randomly to plan 1 or plan 2. The dietitian records the weight losses over a six-week period for participants following plan 1 and those following plan 2.

Instructions for conducting the two-sample Kolmogorov-Smirnov procedure using SPSS are as follows:

Analyze – Nonparametric Tests – Legacy Dialogs – Two Independent Samples Test – Move the Dependent Variable, weightloss, to the Test Variable list – Move plan to the Grouping Variable box – Click Define

Frequencies

	Plan	N
Weightloss	Plan 1	5
	Plan 2	5
	Total	10

Test Statistics[a]

		Weightloss
Most Extreme Differences	Absolute	0.600
	Positive	0.600
	Negative	0.000
Kolmogorov-Smirnov Z		0.949
Asymp. Sig. (2-tailed)		0.329
Exact Sig. (2-tailed)		0.286
Point Probability		0.206

[a] Grouping Variable: plan

FIGURE 4.20
Results of Kolmogorov-Smirnov test for two independent samples.

Groups – Type 1 for plan 1 and type 2 for plan 2 – Click Continue – Unclick Mann-Whitney U Test and click Kolmogorov-Smirnov Z – Click the Exact button – Click Exact – Click Continue – Click OK

Results of the K-S test are not statistically significant: K-S = 0.95, exact *p*-value = 0.29. The dietitian must retain the null hypothesis of no difference in the distributions for the two samples. The dietitian can conclude that the samples came from the same distribution. Results of the K-S test are displayed in Figure 4.20. Since the results of the analysis are not statistically significant, no further analysis, such as mean scores, standard deviation, or median values is necessary.

The Kolmogorov-Smirnov test assesses the difference in the median value of two independent samples. The Hodges-Lehmann estimate for the confidence interval calculates whether the difference in two medians is different from zero. The Hodges-Lehmann test produces an unbiased estimate of the confidence interval within which the difference in the median values lie. Section 4.5 discusses the Hodges-Lehmann estimate for the confidence interval.

4.5 Hodges-Lehmann Estimate for Confidence Interval

A confidence interval is a range of values with the lowest value (lower limit) at one end and the highest value (upper limit) at the other. Confidence intervals allow assessment of the probability that the point estimate (mean or median in a symmetrical distribution) is located within the interval plus or minus the distance from the observed estimate.

For example, a researcher's interest may be in estimating the probability that the true population median is within a certain interval or range of values for a single sample. If a sample yields a median value of 24, the range of values (confidence interval) is one within which the observed value of 24 lies. If one selected 100 random samples from similar populations, one could be confident that the median value for the population would fall within the calculated interval 95 out of 100 times, given an alpha level of 0.05. A 95% confidence interval accounts for 95% of the area under the normal curve; consequently, the probability of observing a value outside the designated interval is 5 or fewer times out of 100.

The researcher sets the boundaries for the confidence interval. The 95% confidence interval is used frequently in behavioral and social sciences research; however, 90% and 99% confidence intervals are also used based on the level of conservatism that the researcher decides is necessary for the specific problem being analyzed.

Some researchers and statisticians support using confidence intervals instead of (or in addition to) *p*-values to ascertain statistical significance because a range of values is produced by confidence intervals.

The Hodges-Lehmann test produces the confidence interval for the difference in the median values of two independent samples that have the same distribution. The only difference between the two groups is their median value. The sample median is the point estimate and the associated estimator is the Hodges-Lehmann. The Hodges-Lehmann is considered an unbiased and most efficient estimator because its sampling variance is minimal.

The test is based on the two end points of the overall data array within which the observed median lies. Hodges-Lehmann estimators are insensitive to distributional qualities of the data, such as normality, skewness, kurtosis, and outliers. However, the data should represent a random sample that is independent and identically distributed (i.i.d.) and also symmetrical.

The Hodges-Lehmann test locates the two end points of the overall data array within which the median lies. The test is based on the calculations of all possible differences between two independent groups. For example, if a group contains 12 cases, then 144 differences can be calculated.

The number of values in one group are multiplied by the number of values in the second group which forms data points comprised of group 1 × group 2 pairs of values; and each pair of data points produces a difference in values. The Hodges-Lehmann is the median value of the 144 differences of the

TABLE 4.6

Paired Differences for Hodges-Lehmann Point Estimate
and Confidence Interval

4.3 – 3.0 = 1.3	4.5 – 3.0 = 1.5	4.9 – 3.0 = 1.9	5.1 – 3.0 = 2.1
4.3 – 3.3 = 1.0	4.5 – 3.3 = 1.2	4.9 – 3.3 = 1.6	5.1 – 3.3 = 1.8
4.3 – 3.5 = 0.8	4.5 – 3.5 = 1.0	4.9 – 3.5 = 1.4	5.1 – 3.5 = 1.6
4.3 – 3.7 = 0.6	4.5 – 3.7 = 0.8	4.9 – 3.7 = 1.2	5.1 – 3.7 = 1.4

paired values. The confidence interval is the smallest difference (lower limit) and largest difference (upper limit) between the pairs. Table 4.6 displays calculations for paired differences between the following data sets: group 1 [4.3, 4.5, 4.9, 5.1] and group 2 [3.0, 3.3, 3.5, 3.7].

Based on the difference values in Table 4.6, the point estimate can be calculated using the formula, $(n_1 \times n_2)/2 = 8$. The sum of $(0.8 + 1.9)/2 = 1.35$; thus, the point estimate is 1.35. The approximate 95% confidence interval for the point estimate is (0.8, 1.9), trimming the extreme values from both the lower and upper ends of the interval. Note that in this illustration other values are smaller than 0.8 and larger than 1.9 (0.6 and 2.1 for the lower and upper limits, respectively). These values are plausible end points for the confidence interval if the extreme values are not trimmed.

The Hodges-Lehmann estimator is the likely range within which the true median lies; therefore, the narrower the range (confidence interval), the more precision one can have in identifying the limits of the true median. If the interval does not overlap zero (does not include zero), one can claim statistical significance and reject the null hypothesis that the difference in the median values between the two groups is equal to zero. In other words, one can conclude that the difference in the medians of the two groups is not equal to zero. However, if zero is in the interval, the null hypothesis must be retained and one cannot claim a difference between the median values. The Hodges-Lehmann test is often conducted along with the Mann-Whitney U test, as illustrated in Example 4.6.

Example 4.6

A mathematics teacher is interested in whether the median scores for boys and girls are different on a pretest for placement in a mathematics study group. The teacher believes that the girls will have an overall higher mean score on the test because they out-performed the boys on a recent science test. She wants to test her hypotheses by first assessing whether differences exist. If she finds statistically significant differences between the scores for boys and girls, she is interested in how confident she can be in the difference in the medians between the groups. The teacher selects test scores randomly from 12 girls and 12 boys for her study and formulates the following research question (RQ) and null hypothesis to test for differences.

RQ: No statistically significant difference exists between the median scores for boys and girls on the mathematics pretest.

H_0: What is the range of values for the difference in the median scores?

First, the teacher conducts a Mann-Whitney U test to assess statistically significant differences between the boys and the girls on a pretest. Recall that the Mann-Whitney U test will reveal whether a difference exists between two groups; however, it does not indicate how large the difference is, that is, the effect size. If the Mann-Whitney U test reveals a statistically significant result, the teacher will follow-up with a test of the Hodges-Lehmann estimate for the confidence interval.

The SPSS path to conduct the Wilcoxon-Mann-Whitey U test is presented in Section 4.3. To conserve space, the steps are not repeated here.

Results of the analysis indicate a statistically significant difference between the median scores for girls and boys (Mann-Whitney U = 23.0, exact p-value < 0.01). The girls scored higher on the pretest than the boys with mean ranks of 16.58 and 8.42, respectively. The results of the Mann-Whitney U test are displayed in Figure 4.21.

Ranks

	Gender	N	Mean Rank	Sum of Ranks
	Female	12	16.58	199.00
Pretest	Male	12	8.42	101.00
	Total	24		

Test Statistics[a]

	Pretest
Mann-Whitney U	23.000
Wilcoxon W	101.000
Z	−2.830
Asymp. Sig. (2-tailed)	0.005
Exact Sig. [2 × (1-tailed Sig.)]	0.004[b]
Exact Sig. (2-tailed)	0.003
Exact Sig. (1-tailed)	0.002
Point Probability	0.000

[a] Grouping Variable: gender

[b] Not corrected for ties.

FIGURE 4.21
Results of Mann-Whitney U test for pretest scores for girls and boys.

Confidence Interval Summary

Confidence Interval Type	Parameter	Estimate	95% Confidence Interval	
			Lower	Upper
Independent-Samples Hodges-Lehmann Median Difference	Differene between medians of pretest across categories of gender.	22.000	9.000	35.000

FIGURE 4.22
Results of Hodges-Lehmann estimate of confidence interval.

The teacher follows up on the significant Mann-Whitney U results by conducting the Hodges-Lehmann test for median difference. Figures 4.6 through 4.9 in Section 4.2 for the median test show the SPSS path and relevant dialog boxes to access the Hodges-Lehmann estimate for the confidence interval. The SPSS path to conduct the Hodges-Lehmann estimate for the confidence interval is as follows:

Analyze – Nonparametric Tests – Independent Samples – Click Customize analysis on the Objective tab – Click the Fields Tab – move the Dependent Variable, Test 1 to the Test Fields box – Move gender to the Grouping Variable box – Click the Settings tab – Click Customize Tests – In the Estimate confidence Interval across Groups section, click Hodges-Lehmann estimate (2 samples) – Click Run

Results show that the difference between the medians yields a point estimate of 22.0. The 95% confidence interval (9.0, 35.0) suggests a fairly wide confidence interval. Note that zero is not in the interval, indicating that the difference between the median scores is statistically significant. Figure 4.22 shows the Hodges-Lehmann results for the point estimate and the confidence interval.

4.6 Moses Extreme Reaction Test

Unlike the previous tests for two independent samples that evaluate differences in the median values of two groups, the Moses extreme reaction test determines differences between two independent groups based on extreme differences in one group. The Moses test may be invoked when other nonparametric tests of median differences such as the Mann-Whitney test or the median test are used; however, the Moses test makes no assumption of equal medians for the groups. In other words, rather than testing for the midpoint (median) in a distribution, the Moses test identifies extreme scores in the tails of a continuous distribution.

The Moses test is used for experimental studies and studies where data are known to be clustered at each end of the distribution rather than at the midpoint. In other words, it is useful when a researcher has reason to believe that the data may be skewed. The Moses test is especially useful for clinical studies such as those involving reactions to medical conditions, treatments, or physiological effects on individuals under different conditions such as fear, anxiety, and happiness. The test can be applied also to behavioral situations in studies for determining participant reactions to changes. Procedures to calculate the Moses test are quite simple; however, hand calculations with large sample sizes are cumbersome.

A short summary of the calculations involved to prepare data for the Moses extreme reaction test is as follows. First, all of the data are ranked, as in the Mann-Whitney U test, with 1 assigned to the lowest rank, and the highest rank assigned to the number of scores in the pooled set of scores. Averaged ranks are assigned for tied scores. The range or span of the ranks is the difference between the largest rank and smallest rank plus 1 and rounded to the nearest integer for one of the groups (usually the control group). All ranks between the smallest and largest ranks are included in the span. The Moses test is illustrated in Example 4.7.

Example 4.7

A statistics professor assigned a class of 17 students randomly to one of two groups (Group A and Group B). Eight students were assigned to Group A, and nine students were assigned to Group B. All students were in the same classroom and received the same instruction at the same time. The only difference between the two groups was the manner of end-of-course grading. Grades for Group A were based on students' participation in class (responding to professor's questions and asking relevant questions). Grades for Group B were based solely on students' mid-term and final examination test scores. Students in Group A were given opportunities to opt out of the participation-only group. No students opted out. At the end of the course, the professor administered a 20-point, end-of-course feedback inventory on which students indicated their reactions to their respective group assignments. The scores and ranks of the scores for the groups are displayed in Table 4.7. The first eight scores are for Group A; scores for cases 9–17 are for Group B.

The professor is interested in testing the null hypothesis that extreme values are as likely in one group as in the other. The null hypothesis is stated as:

H_0: No statistically significant difference in the dispersion of scores for the students in the class participation only group and those in the test only group.

Simply stated, one may write the null hypothesis as:

H_0: The populations are dispersed similarly.

TABLE 4.7

Rank Calculations for Moses Extreme
Reaction Test

Case Number	Group Value	Rank
1	6	5
2	7	6
3	9	7
4	10	8
5	12	10
6	13	11
7	14	12
8	16	14
9	1	1
10	2	2
11	4	3
12	5	4
13	11	9
14	15	13
15	18	15
16	19	16
17	20	17

The alternative hypothesis may be written as:.

H_A: The populations are dispersed differently.

If the p-value is < 0.05, one must reject the null hypothesis and conclude that extreme values are not equally likely between the two groups.

The SPSS path to conduct the Moses extreme reaction test is the same as the path for the median test presented in Section 4.2, with the exception of selecting the Moses extreme reaction (2 samples) in the Compare Ranges across Groups panel of the dialog box displayed in Figure 4.9.

Figure 4.6 in Section 4.2 for the median test shows the path to the Moses extreme reaction test and Figures 4.7 through 4.9 show important dialog boxes to access the test; therefore, the steps of the procedure are not repeated here.

Results of the analysis to test for a statistically significant difference between the two groups of students show a statistically significant difference in the dispersion of scores ($p = 0.043$); therefore, the professor must reject the null hypothesis that the scores are dispersed equally for the groups. Figure 4.23 shows these results.

The box plots and statistical information displayed in Figure 4.24 reveal that the test-only group ("experimental group" in the figure) had more extreme scores than the participation-only (control) group. Note that the span of the participation group is 10.0, which can be assessed easily from Table 4.7, and the one-sided *p*-value is 0.012. Usually, one would interpret the

Hypothesis Test Summary

	Null Hypothesis	Test	Sig.	Decision
1	The range of score is the same across categories of group.	Independent-Samples Moses Test of Extreme Reaction	0.043[a]	Reject the null hypothesis.

Asymptotic significances are displayed. The significance level is 0.05.

[a] Exact significance is displayed for this test.

FIGURE 4.23
Results of Moses extreme reaction test.

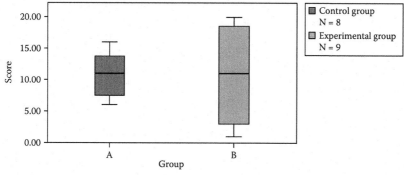

Independent-Samples Moses Test of Extreme Reaction

Control group N = 8
Experimental group N = 9

Total N		17
Observed Control Group	Test Statistic[a]	10.000
	Exact Sig. (1-sided test)	0.012
Trimmed Control Group	Test Statistic[a]	7.000
	Exact Sig. (1-sided test)	0.043
Outliers Trimmed from each End		1.000

[a] The test statistic is the span.

FIGURE 4.24
Box plots and statistical information for the Moses extreme reaction test.

one-sided *p*-value, especially when one group is expected to show extreme responses. As noted in Figure 4.24, extreme values (outliers) are excluded from the analysis. Usually, outliers are values at the lowest and highest 5% of the data array.

Chapter Summary

Fisher's exact test for 2 × 2 tables is used to assess differences between two discrete dichotomous variables. Even though interest is in Fisher's exact test, recall that the chi-square procedure in the Crosstabs menu is used. Fisher's exact test is interpreted from the SPSS results and is the appropriate statistic to interpret with small sample sizes and with any expected cell size less than 5.

The median test is based on finding the median for all of the values and comparing each value in the data set to the median. A value equal to or less than the median is classified in one group, and values greater than the median are classified into the second group.

The Wilcoxon-Mann-Whitney U test examines whether the ranks of one group are higher or lower than the ranks of a second group.

The Kolmogorov-Smirnov test examines the distribution of two samples to assess the difference between the distributions of two populations. The null hypothesis is that the population distributions are the same. If the *p*-value is statistically significant, the null hypothesis must be rejected, and one can claim that the two distributions came from different population distributions. If the null hypothesis is retained, one can conclude that the two samples are from the same population distribution.

The Hodges-Lehmann test estimates the confidence interval for the difference between the medians of two independent samples. The test produces the confidence interval for the difference in the medians of two groups. Specifically, the difference between every possible pair of scores is calculated, and the differences are ranked in ascending order. The median rank is the midpoint in the array of ranks. The Hodges-Lehmann test is considered to be the best unbiased estimate of the median for two independent samples.

The purpose of the Moses extreme reaction test is to determine whether a statistically significant difference exists between the distributions of two independent groups based on extreme values for one of the groups. Although the Moses procedure ranks the data, the raw data should be on at least an interval scale. The basic premise of the Moses test is that extreme scores are equally likely for either group.

Student Exercises

Data files for these exercises can be retrieved online from the file Chapter4data.zip under the Downloads tab at http://www.crcpress.com/product/isbn/9781466507609.

4.1 A nutritionist hypothesizes that a greater proportion of men than women preferred coffee more often than tea as their beverage of choice when given the option. The nutritionist is interested in whether an observed difference is statistically significant.

a. Conduct the appropriate test to assess the association between the two classifications: gender and drink.

b. Evaluate and report the probability of observing cell counts as extreme or more extreme than those observed.

c. Create a bar chart showing beverage of choice by gender.

4.2 A librarian in an elementary school believes that more fiction books are checked out by girls than boys. The librarian keeps a record of the number of books about animals, adventure, romance, suspense, and biographies that were checked out by all students in two fourth grade classes for a three-week period. Perform the appropriate analysis to assess whether there was a difference in the median number of fiction books checked out by girls and boys.

a. Report the median number of books checked out over the three-week period.

b. Was there a statistically significant difference in the number of fiction books checked out by girls and boys? Provide the evidence for your response.

c. How many girls checked out more than the median number of books? How many boys checked out fewer than the median number of books?

4.3. A college golf coach wants to improve the pitching ability of beginning golfers. The coach decides to design a practice program specifically to improve pitching. If the program improves pitching ability, the coach will implement the program as part of the golf team's required conditioning. The coach recruits 50 entering freshmen who are non-golfers. Each of the recruits agrees to demonstrate his pitching ability. Thirty-five of the freshmen demonstrated consistent pitching ability. From the 35 recruits, the coach selected 20 students randomly for the study. Ten of the 20 freshmen were assigned to an experimental group randomly and the remaining 10 served as a control group. The coach provided three weeks of training in pitching for the experimental group. The control group did not receive the training. At the end of the training, the experimental and control groups demonstrated their pitching ability. Each recruit was given five opportunities to pitch. The coach recorded the average number of feet that the ball landed from the flag for each player. Perform the appropriate analysis and respond to the following questions.

a. Is there a statistically significant difference between the median values of the two groups? How do you know?

b. Which group had the higher mean rank? Interpret the meaning of the higher mean rank.

c. What should the coach conclude about the training program?

d. Create a bar chart that displays performance for each of the groups.

4.4 A fifth-grade teacher wants to compare fifth-grade students' memories of multiplication tables at Fairview School and at Ridgeway School. She selects ten students randomly from Fairview and from Ridgeway for her study. The teacher wants to determine the extent to which there is a difference in the percentage of errors between fifth graders at Fairview and those at Ridgeway on a verbal recitation test of multiplication tables. The teacher's hypothesis is that Ridgeway students would have greater memory (fewer errors) on the multiplication tables than students at Fairview because students at Ridgeway were recently exposed to the tables. The null hypothesis is: there is no statistically significant difference in the proportion of errors made in reciting multiplication tables between students in the fifth grade at Fairview and their counterparts at Ridgeway. Her alternative hypothesis is Fairview students make proportionately fewer errors than Ridgeway students in reciting the multiplication tables.

a. Conduct the appropriate statistical procedure to test the teacher's hypothesis and interpret the results.

b. What should the teacher conclude about the proportion of errors made by students at Fairview and those at Ridgeway?

4.5 The chief executive officer of a development office at a large university is interested in the levels of job satisfaction for employees who interact with potential donors and those who have no interactions with potential donors. The CEO selects eight employees randomly who interact with potential donors and seven employees who have no interactions with potential donors. Employees indicate their levels of job satisfaction on a scale that ranges from 0 for absolutely no job satisfaction to 25 for the highest level of job satisfaction. Responses are anonymous. The CEO wishes to test the null hypothesis of group differences. In addition, he wishes to know how confident he can be of the interval in which a difference, if any, may lie. Perform the appropriate tests and respond to the following items.

a. What is the name of the test you performed? Report the test results.

b. Report your decision whether to reject or retain the null hypothesis and write your conclusions based on the test you performed.

c. Conduct a Hodges-Lehmann estimate for the confidence interval. Report the limits of the confidence interval.

d. Do the results of the Hodges-Lehmann estimate support the decision you made about the null hypothesis? Provide the evidence for your response.

4.6 Thirty-five individuals with high triglyceride levels volunteered to participate in an experiment to lower their triglycerides by taking a prescribed medication. Ten participants were selected randomly from this group (medicine only). Another group of 25 individuals with high triglyceride levels did not wish to take prescribed medication; however, they agreed to participate in the experiment by adhering to a diet designed to lower their triglyceride levels. Ten individuals were selected randomly from this group (diet only). Perform the appropriate test to assess whether a difference exists between the groups based on the dispersion of scores, and respond to the following questions.

a. What test did you perform? What were the results?

b. What conclusions can you make based on the results?

4.7 Design a study in an area of interest to you in which you compare differences between two groups. Include the following information in your design.

a. Write a brief scenario of the problem that you wish to investigate.

b. Identify the independent and dependent variables.

c. State the null hypothesis to be tested.

d. Write the name of the test you will use to test the hypothesis.

e. Use a table of random numbers to generate values for the responses for the groups.

f. Conduct the statistical procedure and write the results of the test. Include the decision rule for the null hypothesis.

g. Include a confidence interval in your results and comment on the findings.

h. Create a graph to illustrate your findings.

i. Write the conclusions based on the results.

References

Bhaskar, B. and Habtzghi, D. (2002). Median of the *p*-value under the alternative hypothesis. *American Statistician*, 56, 202–206. http://www.jstor.org/stable/3087299

Coult, A. D. (1965). A note on Fisher's exact test. *Anthropologist*, 67, 1537–1541. http://www.jstor.org/stable/669176

Freidlin, B. and Gastwirth, J. L. (2000). Should the median test be retired from general use? *American Statistician*, 54, 161–164. http://www.jstor.org/stable/2685584

Gibbons, J. D. and Chakraborti, S. (2010). *Nonparametric Statistical Inference*, 5th ed. Boca Raton, FL: Chapman & Hall/CRC.

Giere, R. N. (1972). The significance test controversy: a reader on methodological perspectives. *British Journal for the Philosophy of Science*, 23, 170–181. http://www.jstor.org/stable/686441

Hollander, M. and Wolfe, D. A. (1999). *Nonparametric Statistical Methods*, 2nd ed. New York: John Wiley & Sons.

Kvam, P. H. and Vidakovic, B. (2007). *Nonparametric Statistics with Applications to Science and Engineering*. Hoboken, NJ: John Wiley & Sons.

Liddell, D. (1976). Practical tests of 2 × 2 contingency tables. *Journal of the Royal Statistical Society*. 25, 295–304. http://www.jstor.org/stable/2988087

Pettitt, A. N. and Siskind, V. (1981). Effect of within-sample dependence on the Mann-Whitney-Wilcoxon statistic. *Biometrika*, 68, 437–441

Sprinthall, R. C. (2012). *Basic Statistical Analysis*, 9th ed Boston: Allyn & Bacon.

Stephens, M. A. (1969). Results from the relation between two statistics of Kolmogorov-Smirnov type. *Annals of Mathematical Statistics*, 40, 1833–1837. http://www.jstor.org/stable/2239571

Zaremba, S. K. (1965). Note on the Wilcoxon-Mann-Whitney statistic. *Annals of Mathematical Statistics*, 36, 1058–1060. http://www.jstor.org/stable/2238219

Zelen, M. (1971). The analysis of several 2 × 2 contingency tables. *Biometrika*, 58, 129–137. http://www.jstor.org/stable/2334323

5

Analyses of Multiple Independent Samples

Chapter 4 presented procedures for analyzing independence of two samples. Often in research projects, we wish to compare observations on multiple (k) samples. The nonparametric tests for multiple independent samples presented in this chapter are equivalent to their parametric counterpart, analysis of variance (ANOVA). Recall that an ANOVA research design is used to ascertain differences among two or more groups when the data are interval, ratio, or ranked.

This chapter covers three procedures used to analyze differences among multiple samples. The first procedure discussed is the Kruskal-Wallis one-way analysis of variance by ranks test commonly used to test differences in three or more groups. Next, the extended median test and the Jonckheere-Terpstra test for ordered alternatives are discussed. Formulas and explanations are presented to help conceptualize test calculations.

The SPSS procedures are included to show steps in the conduct of the analysis for each test. Each test is illustrated with an example. The chapter concludes with student exercises to reinforce learning and practice of the tests presented. The data sets for the examples and exercises of this chapter can be retrieved online from the file Chapter5data.zip under the Downloads tab at http://www.crcpress.com/product/isbn/9781466507609.

5.1 Kruskal-Wallis One-Way Analysis of Variance by Ranks Test

The Kruskal-Wallis one-way analysis of variance by ranks (Kruskal-Wallis) is an omnibus test that compares the medians of three or more samples on one dependent variable. This test is the nonparametric counterpart of the parametric one-way analysis of variance (ANOVA) that measures differences in the means of three or more samples. The Kruskal-Wallis test may be thought of as an extension of the Wilcoxon Mann-Whitney U test for measuring differences between two independent groups. Data may be ordinal, ratio, or interval.

The Kruskal-Wallis test may be used when the data are not normally distributed and meeting other ANOVA assumptions is questionable. However, Kruskal-Wallis is almost as effective as the ANOVA procedure when the assumptions of normality and equal variances are met. Although the distributions from which the samples are taken do not need to be normally distributed, the shapes of the distributions should be similar, with a difference only in the medians. In other words, samples may have a positive or negative skew, yet have different medians.

Subjects should be selected randomly and each subject should be independent of all others. Likewise, groups should be independent of all other groups. The design of the study should ensure independence of the error associated with the difference between each value and the group median. Sample sizes should be larger than five cases each, and unequal sample sizes are acceptable.

The Kruskal-Wallis test assesses whether three or more samples come from the same distribution. Differences in the ranks of observations are assessed by combining observations for all groups and assigning ranks from 1 to N to each observation, where N is the total number of observations in the combined array of scores. The lowest value is assigned a rank of 1 and the highest value is assigned a rank corresponding to the total number of values in the combined groups. Tied values are assigned the average of the tied ranks. Ranks in each group are summed; and the sums of the ranks are tested for their differences. A single value is created from the combined differences in the sums of the ranks. The formula with explanations for computing the Kruskal-Wallis H statistic is given in Figure 5.1. The null hypothesis is stated as:

H_0: The population medians are equal.

The alternative hypothesis is stated as:

H_A: At least one of the group medians is different from the other groups.

The following example applies the Kruskal-Wallis test.

Example 5.1

A large high school offers three types of algebra courses: technical algebra, general algebra, and pre-college algebra. All students have

$$H = \frac{12}{N(N+1)} \left(\frac{\sum R_1^2}{n_1} + \frac{\sum R_2^2}{n_2} + \ldots + \frac{\sum R_i^2}{n_i} \right) - 3(N+1)$$

FIGURE 5.1
Formula for Kruskal-Wallis H test. H = Kruskal-Wallis statistic, N = total number of subjects, R_1 to R_i = rankings for Group 1 to last group (i), and n_1 to n_i = number of subjects for Group 1 to last group (i).

access to at least one course, and no student is assigned a specific course. A principal is interested in the type of algebra course that best serves at-risk students identified as performing below grade level. She asks a teacher of the three different types of algebra courses to record the final scores for ten randomly selected at-risk students enrolled in each type of course. None of the students has taken a prior course in algebra. The principal believes that the median scores of all three groups are equal.

The SPSS path for the Kruskal-Wallis test is as follows.

Analyze – Nonparametric Tests – Legacy Dialogs – K Independent Samples – Move score to the Test Variable List box – Move coursetype to the Grouping Variable box – Click Define Range – type 1 in the Minimum box – Type 3 in the Maximum box (for 4 groups, type a 4 in the Maximum box; for 5 groups, type a 5, etc.) – Click Continue – Select Kruskal-Wallis H in the Test Type section of the Tests for Several Independent Samples dialog box – Click Options – Click Descriptive – the default Exclude Cases Test by Test should remain selected – Click Continue – OK

The SPSS dialog boxes for the Kruskal-Wallis H test and the range of grouping variable are displayed in Figure 5.2.

SPSS returns the mean ranks for each of the groups, the Kruskal-Wallis H statistic as a chi-square statistic, degrees of freedom equal to the number of groups minus one, and a *p*-value. The *p*-value tests the probability of obtaining median values as different (far apart) or more so than those observed. The larger the chi-square statistic, the larger the difference in the medians among the rank sums of the groups. The overall Kruskal-Wallis test results are shown in Figure 5.3.

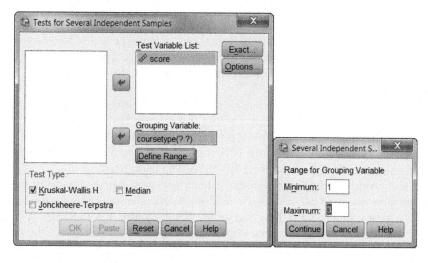

FIGURE 5.2
SPSS dialog boxes for Kruskal-Wallis H test and Range of Grouping Variable.

Ranks

	Coursetype	N	Mean Rank
Score	Technical Algebra	10	16.65
	General Algebra	10	19.75
	Pre-college Algebra	10	10.10
	Total	30	

Test Statistics[a,b]

	Score
Chi-Square	6.274
df	2
Asymp. Sig.	0.043

[a] Kruskal Wallis Test
[b] Grouping Variable: coursetype

FIGURE 5.3
Results of Kruskal-Wallis test for scores by type of algebra course.

Interpretation of the omnibus test is based on the p-value. A small p-value indicates that the median for at least one of the groups is different from those of one or more of the other groups. With p-values less than the preset p-value (usually 0.05 or 0.01), follow-up tests should be performed to determine which groups are different. An effect size index for the overall test is not very informative. However, effect size statistics should be calculated for those comparisons that are statistically significant in the follow-up tests.

With only three groups, the least significant difference (LSD) method may be used to control for Type I error across the groups. However, if more than three groups are compared, then a method such as the Bonferroni or Holm's sequential Bonferroni procedure is preferred. If the p-value is not less than the preset p-value, follow-up tests should not be conducted.

Results reported in Figure 5.3 indicate statistically significant differences among the groups [$\chi^2(2, N = 30) = 6.27, p = 0.04$]. The mean ranks for technical algebra, general algebra, and pre-college algebra are 16.65, 19.75, and 10.10, respectively. Based on the significant p-value, we must reject the null hypothesis and conclude that there is a significant group effect. We will use the Mann-Whitney U test to conduct pairwise comparisons among the groups. The SPSS path for conducting the Mann-Whitney U test is included in Chapter 4. The path for conducting Mann-Whitney U follow-up tests to a significant Kruskal-Wallis statistic is as follows.

> Click Analyze – Nonparametric Tests – Legacy Dialogs – Click 2 Independent Samples – Mann-Whitney U test should already be selected – Move score to the Test Variable(s) box – Move group to the Grouping Variable box – Type 1 for Group 1 and 2 for Group 2 – (For comparisons between Groups 1 and 3, type a 3 in the Group 2 box; for comparisons between groups 2 and 3, type a 2 in the Group 1 box and 3 in the Group 2 box.)

Whether Group 1 is in the Group 1 box or the Group 2 box does not matter. The group placements are arbitrary.

Results of the Mann-Whitney follow-up tests yielded p-values > 0.05 for a comparison between technical algebra and pre-college algebra and a comparison between technical algebra and general algebra. Only the comparison between general algebra and pre-college algebra was statistically different, Mann-Whitney U = 15.50, Z = –2.61, $p < .01$. These results are displayed in Figure 5.4.

Results suggest that the type of algebra course has an effect on students' final test scores. The mean rank for general algebra was 13.95 (sum of ranks = 139.50), and the mean rank for pre-college algebra was 7.05 (sum of ranks = 70.50). The effect size ($r = -2.610/\sqrt{(20)} = -0.58$), which is a large effect

Ranks

	Coursetype	N	Mean Rank	Sum of Ranks
	General Algebra	10	13.95	139.50
Score	Pre-college Algebra	10	7.05	70.50
	Total	20		

Test Statistics[a]

	Score
Mann-Whitney U	15.500
Wilcoxon W	70.500
Z	–2.610
Asymp. Sig. (2-tailed)	0.009
Exact Sig. [2 × (1-tailed Sig.)]	0.007[b]

[a] Grouping Variable: coursetype
[b] Not corrected for ties.

FIGURE 5.4
Results of Mann-Whitney U follow-up test for general algebra and pre-college algebra.

size based on Cohen's estimates. The sign of r provides no useful information, so one would report the absolute value.

Students in the general algebra course had a higher median score (77.00) than those in the pre-college algebra course (62.50). The median score for students in the technical algebra course was 71.00. The principal and teachers may make advisement and curricular decisions based on these results or continue analyzing similar data to confirm these findings.

One advantage of the Kruskal-Wallis test over the median test is that the Kruskal-Wallis test uses more information from the data. It assesses the difference between the mean ranks of the samples, whereas the median test is concerned only with observations at or below the median and those above the median. The extended median test is discussed in Section 5.2.

5.2 Extended Median Test

The median test, also known as Mood's median test, for two independent samples as discussed in Chapter 4 can be extended to three or more samples. It compares the medians of multiple independent samples for data measured on an ordinal scale or higher. Like the median test for two independent samples, an advantage of the median test extended to multiple samples is its insensitivity to departures of homogeneity of variance. No assumptions of data distributions across groups need to be met except that distributions should have similar shapes. The test does require that samples be drawn randomly from their populations. Independence within and between samples is also required.

The test combines all observations into one list while retaining group membership. Observations are then ranked in ascending order, and the grand median is calculated. Basically, the test assesses the frequency of observations that fall at or below the median and the frequency of observations that fall above the median by group. The median used for comparison is for the pooled samples. The data may be cast into a $2 \times c$ table where 2 is the number of rows (one row for frequencies at or below the median and one row for frequencies above the median) and c is the number of columns (one for each variable).

The median test is computed as a chi-square statistic with degrees of freedom equal to (rows − 1) × (columns − 1). If the chi-square statistic is larger than that required to meet the preset alpha level (usually 0.05 or 0.01), we must reject the null hypothesis and declare that the population medians are not equal. The null hypothesis is stated as:

H_0: No difference exists in the population medians from which the samples were taken.

The non-directional alternative hypothesis is stated as:

H_A: At least one of the sample medians is different from one or more others.

Example 5.2 illustrates use of the procedure.

Example 5.2

A manager of a college bookstore is interested in whether there is a difference in the median number of books sold back to the book store at the end of the semester by discipline. He decides to record the number of books sold back for three different senior-level courses: literature, history, and philosophy. He counts the number of books sold back in each discipline and records the numbers in a table. The SPSS path for conduct of the median test is as follows.

Click Analyze – Nonparametric Tests – Legacy Dialogs – Click K Independent Samples – Unclick Kruskal-Wallis and Click Median in the Test Type section – Move booksold to the Test Variable List box – Move subject to the Grouping Variable box – Click Define Range – Type 1 in the Minimum Value box – Type 3 in the Maximum Value box – Click Exact – Select Exact – Click Continue – OK

Results show a statistically significant subject effect for number of books sold back to the bookstore [$\chi^2(2, N = 45) = 11.67, p < 0.01$]. The number of books sold back by discipline varied from equal to or below the median to above the median. The results indicate that follow-up tests are needed to determine which disciplines had more or fewer books sold back that were equal to or below the median and above the median. The results are displayed in Figure 5.5.

Conducting pairwise comparisons for a significant median test requires creating a dichotomous variable from the ranked scores and then conducting follow-up chi-square tests on the new variable. Do not conduct the pairwise comparison follow-up tests within the Tests for Several Independent Samples dialog box we used to conduct the overall median test.

Consider three groups: group1, group2, and group3. When we conduct the overall median test, minimum and maximum group ranges are required. Therefore, input 1 for minimum and 3 for maximum. The follow-up tests for group1 and group2 and for group2 and group3 could be entered as minimum and maximum values. However, the comparison of group1 and group3 could not be made with the overall Median test because all groups would be included when 1 is entered as minimum and 3 is entered as maximum. In addition, there is an added problem of a different median being used for each set of comparisons, risking inaccurate results. We can use the following

Frequencies

		Subject		
		Literature	History	Philosophy
Booksold	> Median	1	10	7
	≤ Median	14	5	8

Test Statistics[a]

	Booksold
N	45
Median	5.00
Chi-Square	11.667[b]
df	2
Asymp. Sig.	0.003
Exact Sig.	0.004
Point Probability	0.003

[a] Grouping Variable: subject
[b] 0 cells (0.0%) have expected frequencies less than 5. The minimum expected cell frequency is 6.0.

FIGURE 5.5
Results of median test for number of books sold back to bookstore.

procedure to rank the cases and compute a median value that will be used by all comparisons. The SPSS procedures to conduct these comparisons are presented in the following four sets of procedures to facilitate ease of application in the somewhat lengthy process. First we will rank the cases to obtain a common median.

Set 1: Rank cases
Click Transform – Rank Cases – Move booksold to the Variable(s) box – Check that Assign Rank 1 to Smallest Value box is selected – Unclick Display Summary Table – Note that there is a button to click for tied scores – the Ties default is the Mean, so leave the default selected – Click OK.
The SPSS dialog box for Rank Cases is shown in Figure 5.6. When you return to the SPSS Data View, you will see a new variable (Rbooksol). You can now create the dichotomous variable as shown in Set 2.

Set 2: Create dichotomous variable
Click Transform – Click Recode into Different Variables – Move Rbooksol to the Numeric Variable – Output Variable box – Type recodebooksol in

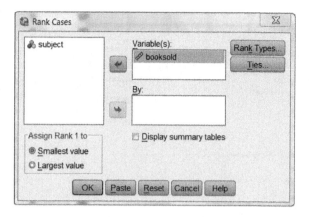

FIGURE 5.6
SPSS dialog box for Rank Cases.

the Name box for the Output Variable – Click Change – Click Old and New Values – click Range – Type 1 in the box below Range and 23 in the Through box (N = 45; (45 + 1)/2 = 23 is the midpoint of the distribution) – Type 1 in the Value box in the New Value section – Click Add – Click All other values in the Old Value section – Type 2 in the Value box in the New Value section – Click Add – Continue – OK

The SPSS dialog box for Recode into Different Variables: Old and New Values is shown in Figure 5.7. When you return to the Data View window, you will see the new recoded variable with values of 1.00 and 2.00. You are now ready to select the cases for the pairwise comparisons as directed in Set 3.

Set 3: Select cases for pairwise comparisons
Click Data – Select Cases – If Condition is satisfied – If – Type subject = 1 or subject = 2 in the Select Cases If dialog box – Click Continue – OK The dialog box that shows cases selected is shown in Figure 5.8.

Returning to the Data View, you will see that only the cases in categories 1 and 2 are allowed in the analysis. The cases in category 3 are given 0 values and a diagonal line appears across the case number. Next, the chi-square comparisons can be made for subjects 1 and 2 (literature and history, respectively). The Select Cases step must be repeated for each set of comparisons, changing the comparison group numbers for each comparison. For example, the next comparison can be between subjects 1 (literature) and 3 (philosophy); and the last comparison will be between subjects 2 (history) and 3 (philosophy).

Set 4: Conduct chi-square follow-up tests
Click Analyze – Descriptive Statistics – Crosstabs – Move recode-booksol to the Row(s) box – Move subject to the Column(s) box – Click

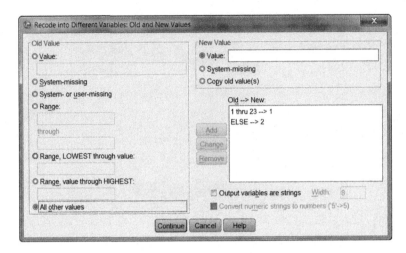

FIGURE 5.7
SPSS dialog box for Recode into Different Variables.

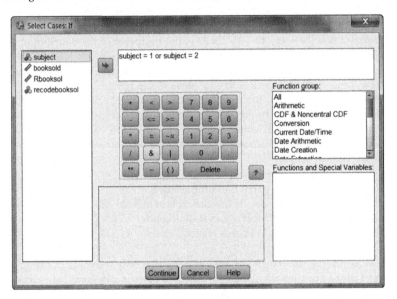

FIGURE 5.8
SPSS dialog box for Select Cases.

Statistics – Chi-square – Click Phi and Cramer's *V* – Click Continue – Click Cells – (Observed should already be selected) – Click Expected – Continue – OK

Results show a statistically significant difference between the number of books sold back to the bookstore for literature and history [$\chi^2(1, N = 30) = 22.53$, $p < 0.01$]. The phi and Cramer's *V* statistics (0.87) show a very

Recodebooksol–subject Crosstabulation

			Subject		Total
			Literature	History	
Recodebooksol	1.00	Count	14	1	15
		Expected Count	7.5	7.5	15.0
	2.00	Count	1	14	15
		Expected Count	7.5	7.5	15.0
Total		Count	15	15	30
		Expected Count	15.0	15.0	30.0

Chi-Square Tests

	Value	df	Asymp. Sig. (2-sided)	Exact Sig. (2-sided)	Exact Sig. (1-sided)
Pearson Chi-Square	22.533[a]	1	0.000		
Continuity Correction[b]	19.200	1	0.000		
Likelihood Ratio	26.893	1	0.000		
Fisher's Exact Test				0.000	0.000
Linear-by-Linear Association	21.782	1	0.000		
N of Valid Cases	30				

[a] 0 cells (0.0%) have expected count less than 5. The minimum expected count is 7.50.

[b] Computed only for a 2 × 2 table

Symmetric Measures

		Value	Approx. Sig.
Nominal by Nominal	Phi	0.867	0.000
	Cramer's V	0.867	0.000
N of Valid Cases		30	

FIGURE 5.9
Results of median follow-up test for literature and history.

large effect size, indicating a strong association between the median number of books sold back and discipline. Figure 5.9 shows results of the follow-up test.

The comparison between literature and philosophy also produced a significant chi-square result [$\chi^2(1, N = 30) = 11.63$, $p < 0.01$]. The phi and Cramer's V statistics (0.62) show a large effect size. The comparison between history and philosophy yielded a non-significant chi-square statistic [$\chi^2(1, N = 30) = 3.33$, $p = 0.07$]. Results for the literature and philosophy comparison and the history and philosophy comparison are not displayed.

Like the Kruskal-Wallis test, the median test is an appropriate statistical procedure when a researcher has no a priori information about the ordering of the scores. The median test counts the number of scores at or below the median or above the median. No hypothesis is formed about the order of the scores. The Jonckheere-Terpstra (J-T) test presented in Section 5.3 addresses analysis of scores when there is a possibility that the scores fall in some order and the order of the scores is of interest.

5.3 Jonckheere-Terpstra Test with Ordered Alternatives

The Jonckheere-Terpstra (J-T) test is also known as the Jonckheere trend test. Its purpose is to examine the medians of two or more populations in which the treatments or conditions are ordered and the data are not distributed normally. One may be tempted to use the Kruskal-Wallis test for differences rather than the J-T test for trends when several independent samples are involved. However, the Kruskal-Wallis (K-W) test compromises power if the medians are ordered and the distributions are non-normal.

The K-W test may fail to detect a difference in the distributions when one exists, whereas the J-T test is more powerful in detecting a trend among responses. The J-T test is commonly used in scientific and medical research such as treatment–response studies, drug reactions with changing dosages, recovery rates for various treatment methods, pulse and heart rate changes with increasing or decreasing temperatures or medications, and strengths and tolerances of materials under different conditions. For example, Richard Feynman, the famous Nobel prize-winning physicist, illustrated the relationship between low temperatures at take-off and O-ring failure as contributing factors to the explosion of the *Challenger* space shuttle in 1986. In addition, the test is useful for studies in education, kinesiology, psychology, sociology, and rehabilitation.

The J-T test uses the medians as location parameters with a natural ordering of the medians. In other words, the median of one population sample is less than the median of the second population sample, and the median of the second population sample is less than the median of the third population sample, etc. The order can be in the opposite direction as well; the median of the first population sample is greater than the median of the second population sample, etc.

The J-T test is the sum of the number of times an observation in each group comes before observations in each of the other groups. The formula is shown in Figure 5.10.

Comparisons are made for individual scores on all combinations of paired groups. For example, a study with three groups (each representing a population), compares Group 1 observations with Group 2 observations at each data point, and a total score is calculated for the comparison. The total score is derived by summing the ranks of one group compared with a second group. Next, observations for Group 1 are compared with those of Group 3 for each

$$\sum_{i=1}^{c-1}\sum_{j=i-1}^{c}\theta_{ij}$$

FIGURE 5.10
Formula for Jonckheere-Terpstra test. Scores in each column are ranked. Each score is compared to every other score in every other column and the number of times the score is less than the comparison score are summed for each column.

case, and a total score is derived for the comparison of Group 1 with Group 3. Finally each score for Group 2 is compared with each score for Group 3 and a final score is calculated for the Group 2–Group 3 comparison.

Each comparison involves finding the number of values in one group (Group 2) that are higher than (or tied with) each value in the other group (Group 1) and assigning a rank to each value in the first group based on the number of scores in the comparison group that are greater than (or tied with) that score. That is, ranks are assigned by considering each value in one group and counting the number of values in the comparison group that are greater than or tied with that value. For tied values, the count is divided between the tied values, such that the count is increased by one half for each group.

The null hypothesis assumes that the groups are from the same population. This assumption means there is no trend in the observations. That is, the distribution is the same across groups. The null hypothesis is stated as:

H$_0$: No difference exists in the order of the mean ranks (medians) of the populations (groups).

The alternative hypothesis is stated as:

H$_A$: The population medians are ordered such that the medians are in an increasing (or decreasing) order.

The researcher should state the alternative hypothesis as medians being in one direction or the other, but not both, unless there is reason to believe that scores will go in one direction to some point and then shift direction. The following example illustrates application of the J-T and follow-up procedures for a study in education.

Example 5.3

A second grade reading teacher believes that students' reading comprehension is greater when students have the usual classroom distractions, rather than absolutely no background noise or a minimal amount of noise such as soft background music. The teacher considers the usual amount of background noise to consist of students talking in groups, teacher talk, and outside distractions.

She tests her hypothesis by randomly assigning 7 students to a no-distraction group (Group 1), 7 students to a minimal-distraction group

(Group 2), and 7 students to a usual-background-noise group (Group 3). The Jonckheere-Terpstra test is applied to examine whether there is a trend in the data. The SPSS path to perform the Jonckheere-Terpstra test is as follows.

Analyze – Nonparametric – Legacy Dialogs – K Independent Samples – Move readcomp to Test Variable box – Move group to Grouping Variable box – Define Range – Type 1 in the Minimum box and type 3 in the Maximum box – Click Continue – Unclick Kruskal-Wallis H – Click Jonckheere-Terpstra – Click Exact – Select Exact – Click Continue – Click OK

Results of the J-T test show that the medians for reading comprehension are not equal across the groups (J-T = 131.00, $p < 0.01$). These results are displayed in Figure 5.11. The teacher interprets the one-tailed test ($p < 0.01$) because she believes the median values are ordered and that there is a trend in the data. Therefore, the null hypothesis is rejected, and the teacher concludes that the median values are different across groups.

Follow-up tests are needed for a significant J-T statistic to examine the trend in the medians. The following procedure will show the trend in the data and produce an effect size.

Analyze – Correlate – Bivariate – Move group and readcomp to the Variables box – Select Kendall's tau-b – Pearson, the default in the Correlation Coefficients section should remain selected – Be sure that

Jonckheere-Terpstra Test[a]

	Readcomp
Number of Levels in group	3
N	21
Observed J-T Statistic	131.000
Mean J-T Statistic	73.500
Std. Deviation of J-T Statistic	15.521
Std. J-T Statistic	3.705
Asymp. Sig. (2-tailed)	0.000
Exact Sig. (2-tailed)	0.000
Exact Sig. (1-tailed)	0.000
Point Probability	0.000

[a] Grouping Variable: group

FIGURE 5.11
Results of Jonckheere-Terpstra test for reading comprehension.

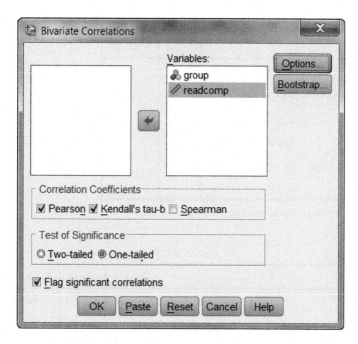

FIGURE 5.12
SPSS dialog box for Bivariate Correlations.

Flag Significant Correlations is selected – Select One-tailed for the Test of Significance – Click OK

The SPSS dialog box with the correct selections to conduct bivariate correlations is shown in Figure 5.12.

Kendall's $\tau_b = 0.66$, $p < 0.01$ indicates a strong association between reading comprehension scores and level of distraction. Figure 5.13 shows those results to be identical to the J-T results. This is always the case. In fact, some researchers may skip the J-T test altogether and simply conduct the Kendall's τ_b test. We can see that the correlation coefficient (0.75, $p < 0.01$) is statistically significant (one-tailed test), indicating a linear trend in the medians of the groups' reading comprehension scores. Based on the evidence, the teacher must reject the null hypothesis of no trend in scores on reading comprehension, and conclude that there is a trend in the scores.

To examine the direction of the trend, we will compute the median values on reading comprehension for each group. The following procedure yields medians by group.

Analyze – Descriptive Statistics – Explore – Move readcomp to Dependent List box – Move group to the Factor List box – Click the Statistics button to check that Descriptives is selected – Click Percentiles – Click Continue – Select Both in the Display section – Click OK

Correlations

		Group	Readcomp
Group	Pearson Correlation	1	0.750[a]
	Sig. (1-tailed)		0.000
	N	21	21
Readcomp	Pearson Correlation	0.750[a]	1
	Sig. (1-tailed)	0.000	
	N	21	21

[a] Correlation is significant at the 0.01 level (1-tailed).

Correlations

			Group	Readcomp
Kendall's tau$_b$	Group	Correlation Coefficient	1.000	0.655[a]
		Sig. (1-tailed)	.	0.000
		N	21	21
	Readcomp	Correlation Coefficient	0.655[a]	1.000
		Sig. (1-tailed)	0.000	.
		N	21	21

[a] Correlation is significant at the 0.01 level (1-tailed).

FIGURE 5.13
Results of Kendall's τ_b following a significant J-T statistic.

Results of the descriptive statistics confirm a trend in the median test values. The median for Group 1 (no distraction) is 19.0; the Group 2 (minimal distraction) median is 34.0; and the median for Group 3 (usual distraction) is 50.0. The median values alone do not confirm rejection or retention of the null hypothesis. They are, however, necessary to interpret where the trend lies when we have statistically significant J-T results.

The results suggest that as the distraction level approaches its usual level, reading comprehension scores increase. The boxplots displayed in Figure 5.14 portray the trend in median values.

The Jonckheere-Terpstra test is very useful when a researcher is interested in whether there is an order to observations. Ordering of the scores should be determined during the analysis, not prior to the analysis. In other words,

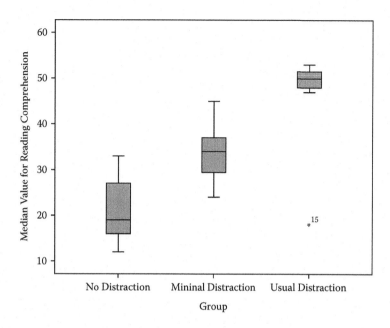

FIGURE 5.14
Boxplots for trend in reading comprehension by distraction level.

one would not explore the data first, decide there is an order, and then conduct the test. The function of the test is to reveal whether an order exists, not to confirm what the researcher has already decided.

Chapter Summary

The Kruskal-Wallis test and the median test are used to measure a single dependent variable for three or more independent samples. The Kruskal-Wallis test is the most popular of nonparametric counterparts to the parametric analysis of variance test. Both tests assess equality of population distributions, and both tests are reported as chi-square statistics. However, the Kruskal-Wallis test is more conservative and shows more sensitivity to the data than the median test.

Both tests are robust to non-normal distributions, as long as the distributions are non-normal in the same way. The median test may be used in a variety of research designs such as comparing median values over time or across ratings. When results of the Kruskal-Wallis or median test show statistically significant differences, follow-up tests are needed to ascertain which groups are different. If the omnibus tests are not statistically significant, no follow-up tests are needed.

The J-T test is used when a researcher is interested in whether observations are ordered in an ascending or descending pattern. The J-T test is based on the order of the values of each group compared to the values of every other group. The total number of times that a particular value is less than a comparison value is summed for each group.

Student Exercises

Data files for these exercises can be retrieved online from the file Chapter5data.zip under the Downloads tab at http://www.crcpress.com/product/isbn/9781466507609.

5.1 A rehabilitation counselor believes that the number of counselor dependency days for clients following rehabilitation increases in order from rotator cuff surgery, hip replacement, and knee replacement. The null hypothesis for this study is that there is no difference in the mean ranks (medians) of the populations for the number of counselor dependency days. Perform the analysis and respond to the following items.

   a. Should the null hypothesis be rejected or retained? Provide the evidence.

   b. Write a conclusion statement including the test statistic and follow-up tests results if appropriate.

   c. Create a plot showing dependency days by condition.

5.2 A sociologist is interested in studying whether environmental concerns of college students vary with their major fields of study. He decides to recruit college seniors from four different major fields of study: business, engineering, liberal arts, and education. Fifteen seniors from each discipline were randomly selected from a larger pool of students for a total of 60 participants. All agreed to respond to a 25-item questionnaire on environmental concerns. Scores on the questionnaire ranged from 25 (very little concern for the environment) to 100 (great concern for the environment). The null hypothesis states no difference among the groups on their concern for the environment.

   a. Conduct the appropriate test for this study, and report the statistic and the p-value.

   b. What is the decision rule applied to the null hypothesis? Should you reject or retain the null hypothesis?

   c. Are follow-up tests needed? Why or why not?

   d. Write a conclusion for this study.

5.3 Forty students, 10 in each of four disciplines (literature, history, science, and philosophy) were selected to enter a public speaking contest on the integration of technology into daily living. All students had good organizational and speaking skills. Judges rated the students on a 20-point scale for originality of content only. Low scores were interpreted to mean not very original and high scores were interpreted to mean highly original. Judges were not informed of the students' disciplines. Test the null hypothesis that all groups of students were taken from a population with the same median.

a. Name the test you would use and perform the procedure.

b. What conclusions can you draw from your test results? Provide the evidence.

c. Create a graph showing scores by discipline.

5.4 A school counselor believes that the number of visits a student makes to the principal's office is related to his or her academic performance. The counselor asks a teacher to provide information for students in one of her classes on the number of their visits to the principal's office and the averages of their last three test scores. The teacher provides the requested information for 19 students.

a. What is the appropriate statistical procedure for this study?

b. Perform the test and report the p-value and the median scores by number of visits to the principal's office.

c. Write a conclusion statement for this problem. Include a graph to illustrate your conclusions.

References

Bhaskar, B. and Habtzghi, D. (2002). Median of the p value under the alternative hypothesis. *American Statistician*, 56 202–206. http://www.jstor.org/stable/3087299

Gibbons, J. D. and Chakraborti, S. (2010). *Nonparametric Statistical Inference*, 5th ed. Boca Raton, FL: Chapman & Hall/CRC.

Hettmansperger, T. P. and Norton, R. M. (1987). Tests for patterned alternatives in k-sample problems. *Journal of the American Statistical Association*, 82, 292–299. http://www.jstor.org/stable/2289166

Hollander, M. and Wolfe, D. A. (1999). *Nonparametric Statistical Methods*, 2nd ed. New York: John Wiley & Sons.

Klotz, J. and Teng, J. (1977). One-way layout for counts and the exact enumeration of the Kruskal-Wallis H distribution with ties. *Journal of the American Statistical Association*, 72, 166–169. http://www.jstor.org/stable/2286931

Levy, K. J. (1979). Pairwise comparisons associated with the K independent sample median test. *American Statistician, 33*, 138–139. http://www.jstor.org/stable/2683817

Lewis, G. H. and Johnson, R. G. (1971). Kendall's coefficient of concordance for sociometric rankings with self excluded. *Sociometry, 34*, 496–503. http://www.jstor.org/stable/2786195

Marden, J. I. and Muyot, M. E. T. (1995). Rank tests for main and interaction effects in analysis of variance. *Journal of the American Statistical Association, 90*, 1388–1398. http://www.jstor.org/stable/2291530

Rayner, J. C. W. and Best, D. J. (2000). *A Contingency Table Approach to Nonparametric Testing*. Boca Raton, FL: Chapman & Hall/CRC.

Simon, G. (1977). A nonparametric test of the total independence based on Kendall's tau. *Biometrika, 64*, 277–282. http://www.jstor.org/stable/2335694

Tanner, D. (2012). *Using Statistics to Make Educational Decisions*. Thousand Oaks, CA: Sage.

Weller, E. A. and Ryan, L. M. (1998). Testing for trend with count data. *Biometrics, 54*, 762–773. http://www.jstor.org/stable/3109782

6

Analyses of Two Dependent Samples

Some research questions are answered best when the samples are paired or dependent upon one another in some way. Tests presented in this chapter are based on cases that represent pairs of scores that are dependent on one another. Dependent samples are also referred to as paired, matched, or related. The terms are used interchangeably.

Dependent samples tests involve research designs in which two of the same or different measures are taken on the same participants or paired samples under two different conditions or at two different times. The purpose of these tests is to ascertain differences between the measures. Unlike their parametric counterparts that focus on inferences from the data, the tests presented in this chapter are concerned with shifts in the median (or mean) values of the paired samples. Participants serve as their own controls. An intervention (treatment) may or may not occur between the first and second measure. Dependent samples designs allow for matching of subjects. One measure is taken on one individual in the pair and another measure is taken on the other individual in the pair; or the same measure can be taken on both individuals in a pair at different times or under different conditions. Each pair serves as a case with two scores. Studies using paired samples may or may not have interventions. Generally, an intervention is applied to both individuals within a pair.

This chapter covers the McNemar change test that assesses changes in responses, the sign test for two related samples to test for differences between two medians, and the Wilcoxon signed rank test for differences in mean ranks between pairs of observations. In addition, the Hodges-Lehmann estimate for the confidence interval for two dependent samples is discussed.

The SPSS procedures are included to show the steps in the conduct of the analysis for each test, and each test is illustrated with an example. A summary is presented at the end of the chapter. The chapter concludes with student exercises to reinforce learning and practice of the tests presented. The data sets for the examples and exercises of this chapter can be retrieved online from the file Chapter6data.zip under the Downloads tab at http://www.crcpress.com/product/isbn/9781466 507609.

6.1 McNemar Change Test

The McNemar change test examines the extent of change in scores from participant responses under one condition or time to a second condition or time, for example, in pretest and posttest situations. An intervening (treatment) measure may or may not be used in the study, depending upon the research design. The design is similar to the parametric procedure that uses a dependent or paired samples t-test. In a paired-subjects design, subjects are matched on a nuisance variable.

Care should be taken that the matching variable is indeed a nuisance variable rather than an extraneous variable. A nuisance variable will not change across levels of the independent variable so that variations in the variable will not systematically affect the outcomes of a study. In other words, matching should be based on a variable that remains constant for levels of the independent variable such as a relationship between a matched pair (father and son; husband and wife; pro-abortion and anti-abortion). On the other hand, an extraneous variable can take on different values for the two levels of the independent variable (scores on achievement tests; incomes). Simply stated, a nuisance variable helps reveal true differences between the two levels of the independent variable. Extraneous variables confound the results of an analysis because values of the dependent variable change with changes in the levels of the independent variable.

In a pretest–posttest design, each subject acts as its own control. Subjects should be selected randomly and their pre- and post scores must be independent of any other pre- and post scores in the sample.

The McNemar test is appropriate for data measured on a dichotomous nominal scale. The test measures changes in responses over time rather than differences in responses. Frequencies of responses recorded in a 2 × 2 table can help reveal the direction in which changes in responses occur.

The directions of participant responses are indicated by plus (+) or minus (–) signs, 0s and 1s in the rows and columns. Yes and no options are also acceptable. Changes in responses are located in the upper left cell from 1 (+) to 0 (–) and in the lower right cell from 0 (–) to 1 (+) of the table. The placement of the 1s (+ symbols) and 0s (– symbols) in the cells in the rows and columns is optional. Changes in the signs will always be on the diagonal. If a 1 (+) is in column 2, then a 0 (–) must be in column 2; likewise, if 1 (+) is in row 1, then a 0 (–) must be in row 2. The proportions to be compared in this design are those in which the frequency of the signs on the diagonal cells changes from 1 (+) to 0 (–) or from 0 (–) to 1 (+). Table 6.1 illustrates this idea.

The null hypothesis is that the number of changes in one direction is just as likely as the number of changes in the other direction. The McNemar procedure tests the null hypothesis of marginal homogeneity. The marginal totals are formed by the sum of frequencies in the cells on the rows and

TABLE 6.1

A 2 × 2 Classification Table for McNemar Test

	(–) y = 0	(+) y = 1	Margin Totals
(+) x = 1	Cell A Frequency of + responses on first measure and – on second measure (1,0)	Cell B Frequency of + responses on first measure and + on second measure (1,1)	Cell A + Cell B
(–) x = 0	Cell C Frequency of – responses on first measure and + on second measure (0,0)	Cell D Frequency of – responses on first measure and – on second measure (0,1)	Cell C + Cell D
Margin totals	Cell A + Cell C	Cell B + Cell D	

$$\chi^2 = \frac{(\text{Cell A frequency} - \text{Cell D frequency})^2}{\text{Cell A frequency} + \text{Cell D frequency}}, \quad df = 1$$

FIGURE 6.1
Chi-square formula to calculate McNemar change test.

the sum of frequencies in the cells on the columns. In other words, since only two cells show the frequency of changed responses, the squared difference in these two cells divided by the sum of the frequency of the two cells will indicate whether the null hypothesis is true. If the null hypothesis is true, one would expect an equal number of responses to change from 1 (+) to 0 (–) as those that change from 0 (–) to 1 (+). The test may be performed with a hand calculator using the chi-square formula presented in Figure 6.1. Example 6.1 illustrates the McNemar test for two related samples (pretest and posttest design with an intervention).

Example 6.1

A statistics professor believes that more college freshmen would elect to take a statistics course if they could participate in a pre-statistics workshop. Her interest is in students' attitudes toward statistics. The professor decides to test her hypothesis that the proportion of students who change their attitudes from agreeing to take a statistics course to not agreeing to take a statistic course is the same as the proportion who change their attitudes from not agreeing to take a statistics course to agreeing to take the course.

The professor arranges for an intensive three-day workshop to introduce students to basic statistical terminology and uses of statistics in daily life. Next, she selects 50 incoming freshmen randomly for her study. Students were asked if they would elect to take a statistics course before participating in the pre-statistics workshop and after participating in the workshop.

The null hypothesis is that the proportion of students who change their minds from agreeing to take a statistics course to not agreeing to take the course is the same as the proportion who change their minds from not agreeing to take a statistics course to agreeing to take the course. The null hypothesis is stated as follows:

H_0: Change in attitude toward statistics is not influenced by the workshop.

The alternative hypothesis is stated as follows:

H_A: There is a change in direction of participant responses.

The frequencies of student responses are displayed in Table 6.2.

The cells in the table are labeled A, B, C, and D for organizational purposes only. When data are recorded in a 2 × 2 table, the weighted cases method can be used to facilitate data entry into the SPSS spreadsheet. Notice that the eight responses in Cell A are labeled 1,1 for students who agreed to take a statistics course before the workshop and those who agreed to take a statistics course after the workshop. The seven responses in Cell D are labeled 0,0 for those who did not agree to take a statistics course before the workshop and those who did not agree to take a statistics course after the workshop. These entries are repetitive and offer no useful information.

Recall that the professor is interested in whether there is a change in the proportion of students who change their minds from agree to not agree or from not agree to agree to take a statistics course after the pre-statistics workshop. Therefore, the focus is on the difference in the proportions in Cells B and C. These cells are labeled 1,0 and 0,1, respectively. The frequency of responses in Cell B is 10, and the frequency in Cell C is 25.

SPSS provides three paths to accessing the McNemar test. The same decision for the null hypothesis will be made regardless of the path used. It is inappropriate and unnecessary to use more than one of the paths for the same study. The decision to reject or retain the null hypothesis will remain the same regardless of the path used. Each path is presented in this chapter for informational and illustration purposes only. The data view for weighted cases and the SPSS dialog box for weight cases appear

TABLE 6.2

Response Frequencies for Pre- and Post-Statistics Workshop

	Post-Statistics Workshop		
	1 = yes	0 = no	Row Totals
Pre-Statistics	Cell A = 8	Cell B = 10	A + B = 18
Workshop	(1, 1)	(1, 0)	
0 = no 1 = yes	Cell C = 25	Cell D = 7	C + D = 32
	(0, 1)	(0, 0)	
Column Totals	A + C = 33	B + D = 17	50

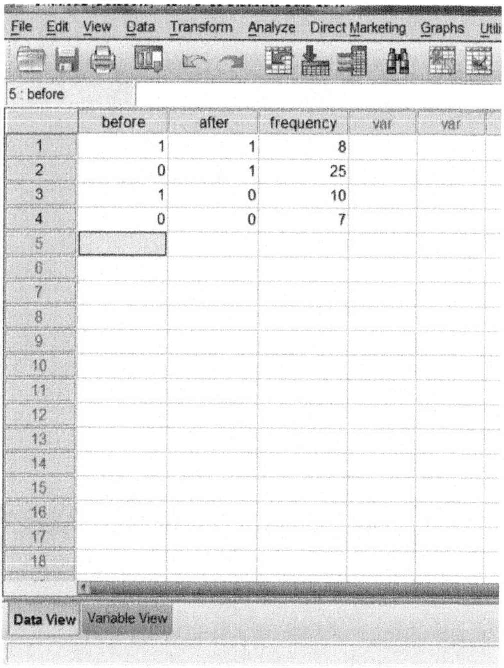

FIGURE 6.2
Weighted cases data for pre- and post-statistics workshop.

in Chapter 3 in Figures 3.3 and 3.4, respectively. For convenience, these relevant dialog boxes are included here. Data from Table 6.2 are shown in Figure 6.2 using the weighted cases method. Figure 6.3 shows the Weight Cases dialog box.

The first SPSS path to access McNemar's test is through the customize analysis method. This path is as follows.

Click Analyze – Click Nonparametric Tests – Click Related Samples – Click the Objective tab – Select Automatically compare observed data to hypothesized in the What is your objective? panel – Click the Fields tab – Move before and after to the Test Fields box – Click Run – (Optionally one could Click the Settings tab, then – Click Customize Test – Select McNemar (2 samples) in the Tests for Change in Binary Data panel – Click Run)

Figure 6.4 displays the SPSS path for related samples. Figure 6.5 displays the dialog box with the Objectives tab open. Notice that the default on the Objective tab automatically compares data to hypothesized data. In the lower panel of the window, various tests including the McNemar are listed. Figure 6.6 shows the Fields tab activated with the appropriate variables selected. The appropriate variables listed should be moved to the Test Fields box. Do not move the frequency variable to the Test Fields box.

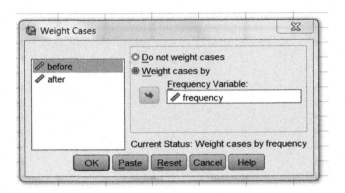

FIGURE 6.3
SPSS dialog box for Weight Cases.

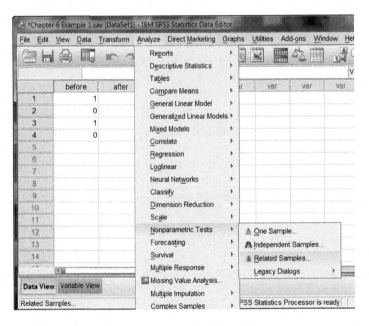

FIGURE 6.4
SPSS path to Related Samples menu.

Figure 6.7 shows the Nonparametric Tests for Two or More Related Samples box with the Settings tab open. Note that Customize tests and McNemar's test (2 samples) have been selected.

Click the Run button on the bottom of the box to see results of the test that include a statement of the null hypothesis that was tested, the name of the test conducted, the *p*-value for the results, and the decision rule. As shown in Figure 6.8, the professor should reject the null hypothesis that changes in

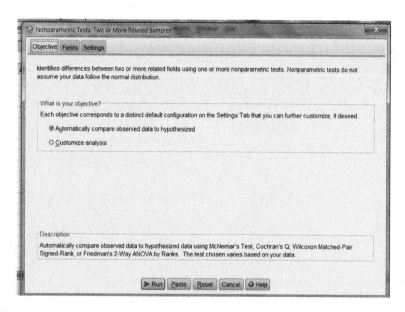

FIGURE 6.5
SPSS dialog box for Nonparametric Tests: Two or More Related Samples with Objective tab open.

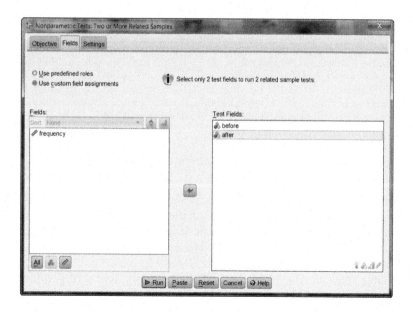

FIGURE 6.6
SPSS dialog box for Two or More Related Samples with Fields tab open; variables in the Test Fields box are shown.

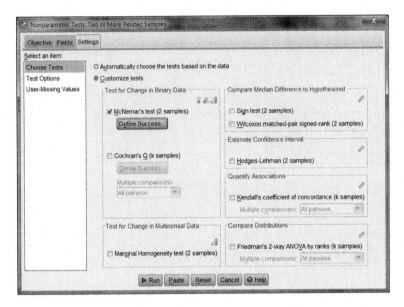

FIGURE 6.7
SPSS dialog box for Two or More Related Samples with Settings tab open and McNemar's test selected.

Hypothesis Test Summary

	Null Hypothesis	Test	Sig.	Decision
1	The distribution of different values across before and after are equally likely.	Related-Samples McNemar Test	0.018	Reject the null hypothesis.

Asymptotic significances are displayed. The significance level is 0.05.

FIGURE 6.8
Hypothesis test summary for McNemar procedure for two related samples.

attitude toward taking a statistics course were not influenced by the workshop $p = 0.02$. A report of the results of the analysis should state the null hypothesis, the name of the test, the probability level, and the decision to reject the null hypothesis. A summary of the results for Example 6.1 is shown in Figure 6.8.

A double click on the Hypothesis Test Summary box will generate bar graphs for the groups and other statistical information. A fourfold table will display a graph for the 0,1 and 1,0 cells. The 0,0 and the 1,1 cells simply report tied responses (same responses for before and after workshop). Note that the test statistic is 5.60. The bar graphs are displayed in Figure 6.9.

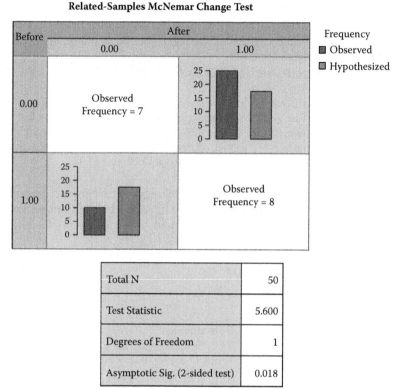

FIGURE 6.9
Test statistic and additional information for results of McNemar test for two related samples.

The McNemar statistic is reported as a chi-square test with 1 df [$\chi^2_{(1)} = 5.60$, $p = 0.02$, $N = 50$]. Under the null hypothesis, the chi-square test of marginal homogeneity is based on a large sample size. If the sum of the frequencies in the compared cells is ≤ 25, the test follows the binomial distribution. If the formula in Figure 6.1 is applied, the test statistic is 6.43 and not 5.60.

This discrepancy between the test statistic calculated using the formula in Figure 6.1 and the test statistic produced by SPSS is due to an adjustment that the SPSS program makes automatically regardless of sample size. The SPSS algorithm adjusts the test statistic down. The adjustment is called the continuity correction—also known as Yates continuity correction. To make the continuity correction, one would need to subtract 0.5 from each of the cell counts where changes in responses are noted (for example, Cells B and C), square the result, and divide by the sum of B and C.

Statisticians usually agree that using the continuity correction should be avoided because it may be too conservative and overcorrect, thus risking a Type II error. Recall that a Type II error is one in which the researcher claims there is no difference, relationship, or effect, when in fact there is. In other words, the researcher commits a Type II error by retaining a false null hypothesis. SPSS invokes the Yates correction automatically. As a point of information and interest, Figure 6.10 displays the formula for the McNemar test with the continuity correction.

The second SPSS path to the McNemar test is as follows.

> Click Analyze – Click Nonparametric Tests – Click Legacy Dialogs – Click 2 Related Samples – Move before to the Variable 1 box in the Test Pairs panel – Move after to the Variable 2 box in the Test Pairs panel – Unclick Wilcoxon if it is checked (Wilcoxon is the default test) – Click McNemar – Click Exact – Select Exact – Click Continue – OK

Figure 6.11 illustrates the SPSS path using the Legacy Dialogs path. Figure 6.12 displays the SPSS dialog box to select the McNemar test. Figure 6.13 shows the SPSS dialog box for selecting the Exact test.

$$C = \frac{(|B-C|-1)^2}{B+C}, \quad df = 1$$

FIGURE 6.10
Formula for McNemar test with continuity correction.

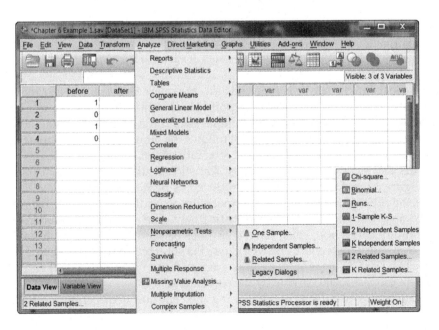

FIGURE 6.11
SPSS path for analysis of two related samples.

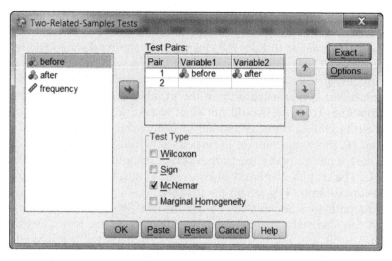

FIGURE 6.12
SPSS dialog box for two related samples tests with McNemar selected.

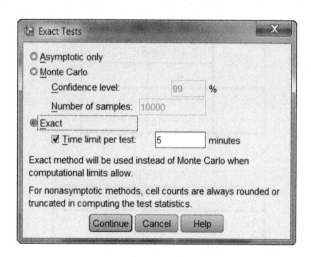

FIGURE 6.13
SPSS dialog box with Exact test selected.

Of course, the results using the Legacy Dialogs path are the same as the results using the Customized analysis path [$\chi^2_{(1)} = 5.60$, $p = 0.02$]. Note that the p-value is based on the asymptotic distribution with the continuity correction. The exact p-value is 0.017. The SPSS output using this procedure also prints a 2 × 2 table. The table shows 15 redundant responses: 8 students agreed to take a statistics course before and also after the pre-statistics workshop (Cell D), and 7 students did not agree to take a statistics workshop before or after the pre-statistics workshop (Cell A). Cell B shows that 25 students changed their minds from not agreeing to take a statistics course before the workshop to agreeing to take the course after the workshop. The ten frequencies in Cell C indicate the number of students who changed their minds from agreeing to take a statistics course to not agreeing after the workshop.

The professor is interested only in the number of students who changed their responses from no (would not elect to take a statistics course) to yes (would elect to take a statistics course); or from yes to no. Consequently, only cells B and C show the changes in responses before and after the workshop. The professor can compare the proportion of responses in Cell B with those in Cell C. The results displayed in Figure 6.14 indicate that the sample proportions are different $[\chi^2_{(1)} = 5.60, p = 0.018]$.

A third path to accessing the McNemar test for two related samples is through the Crosstabs menu. The Crosstabs procedure was illustrated in Chapter 3, so it will not be repeated here. Crosstabs will produce the frequencies and percents of change for the different cells and the *p*-value for the McNemar test. As reported previously in this example, 25 students or 78% (25/32 = 78) changed their pre-workshop responses from no to yes, and 10 students or 56% (10/18 = 56) changed their pre-workshop responses from

Before & After

Before	After	
	0	1
0	7	25
1	10	8

Test Statistics[a]

	Before & After
N	50
Chi-Square[b]	5.600
Asymp. Sig.	0.018
Exact Sig. (2-tailed)	0.017
Exact Sig. (1-tailed)	0.008
Point Probability	0.005

[a] McNemar Test
[b] Continuity Corrected

FIGURE 6.14
Results of McNemar test for Example 6.1.

Before–After Crosstabulation

			After 0	After 1	Total
Before	0	Count	7	25	32
		% within before	21.9%	78.1%	100.0%
	1	Count	10	8	18
		% within before	55.6%	44.4%	100.0%
Total		Count	17	33	50
		% within before	34.0%	66.0%	100.0%

Chi-Square Tests

	Value	Exact Sig. (2-sided)	Exact Sig. (1-sided)	Point Probability
McNemar Test		0.017[a]	0.008[a]	0.005[a]
N of Valid Cases	50			

[a] Binomial distribution used.

FIGURE 6.15
Results of McNemar test using Crosstabs procedure.

yes to no. These results are informative for planning future pre-statistics workshops. The results using the Crosstabs menu are shown in Figure 6.15.

The three different ways to access the McNemar test presented in this chapter are for illustration purposes only. Obviously, one would not use all three ways for the same analysis. Preference for one SPSS path over another is based on personal choice. For example, if a researcher is interested in the hypothesis test summary and bar graphs, the customized analysis is the best choice. The use of signs (+ and –) for the McNemar test to show direction of change is closely linked to the related samples sign test. The sign test for two related samples is discussed in Section 6.2.

6.2 Sign Test for Two Related Samples

The sign test for two related samples is appropriate when researcher interest is whether one random variable in a pair is larger than the other random variable in the pair. This test is similar to its parametric counterpart, the dependent or paired samples t-test. Two measures of the same variable or measures of two different variables of interest are taken at two different times or under two different conditions on the same individual. In each situation, the goal is to assess the change in scores from the first observation to the second observation. This is a repeated measures or pretest–posttest design in which individuals act as their own controls. Specifically, the test ascertains the probability of observing differences between two random variables and evaluates whether the direction of change between two random variables can be attributed to chance.

The two related samples test also accommodates matched or paired samples. For matched pairs, each individual in a pair is tested under the same or different conditions to assess whether a difference exists in the scores of the paired individuals. Matching should be based on a nuisance variable rather than an extraneous variable. A nuisance variable remains constant across levels of the independent variable such as age group, grade level, or gender. An extraneous variable such as test scores varies with individuals, and the variations can systematically influence the results of an analysis. An intervention may be used in either a repeated measures or matched pairs design.

Assumptions for the related samples test require that the bivariate random variables be mutually independent of one another. Individual scores should be independent of one another within groups and between groups. Data should be measured on at least an ordinal scale. Matching within each pair is permitted, so long as the matching criterion does not vary among individuals. No other assumptions about the sample are required.

Like the one-sample test, the sign test for two related samples uses the binomial distribution where the probability of an event occurring is $p = \frac{1}{2}$, and the probability of an event not occurring is equal to $q = p - \frac{1}{2}$ or $1 - \frac{1}{2}$. Quantitative differences are ignored, and only the proportion of differences in scores in one direction or the other is of interest. Tied scores are omitted from the analysis. Dropping tied scores decreases the sample size; however, tied scores do not show differences in proportions in either direction.

The null hypothesis states that there is no difference between measures taken at two different times or under two different conditions for two random variables. The test is appropriate for both experimental and observational studies. The null hypothesis for observational studies is that the mean (or median) difference between pairs is zero; for experimental studies, the null hypothesis is stated as no treatment effect. In other words, the null hypothesis states that the mean (or median) difference between the groups on the first and second observation is equal to zero. The null hypothesis can be rejected when the number of positive (or negative) differences between the pairs is significantly different from 50%. If there is no difference between the pairs of scores, one can expect both groups to have an equal chance (0.50) of a positive difference between each pair of observations. The number of difference scores that may be denoted by m (number out of the total number of observations) follows a binomial distribution for 25 or fewer cases. The one-tailed and two-tailed exact significance (*p*-values) are reported. Since researcher interest is in the probability of observing difference scores in one or the other direction that are far from $n/2$, the one-tailed p-value is interpreted. Sample sizes larger than 25 are based on the Z distribution, and the asymptotic two-tailed p-value is reported. The one-sided p-value is found by dividing the two-tailed p-value by two.

The sign test for related samples is based on the frequency of differences between the numbers of pairs of scores in one group that exceed the numbers of pairs of scores in the other. A change in a positive direction is identified with a plus (+) sign, and a change in a negative direction is identified with a minus (–) sign. If there are no differences between the groups, one expects to observe an equal number of positive and negative score differences in each group. However, the evidence will not support the null hypothesis (the median difference between pairs is zero) if the number of pairs of scores is far from one-half or $n/2$, where n = the number of pairs.

The formula in Figure 6.16 illustrates calculation of the one-sided p-value based on a standard normal distribution. Application of the formula using a *continuity correction* (subtracting ½ in the numerator of the formula) for the data in Table 6.3 is for illustrative purposes only. The continuity correction is an adjustment that improves an approximation to the standard normal distribution for small sample sizes.

Table 6.3 shows 20 pairs of scores; 15 of the pairs have positive differences. The one-sided normal approximation of Z may be calculated using the formula in Figure 6.16. Applying that formula to the data in Table 6.3, the approximate normal Z-statistic is calculated as follows: $Z = 15 - (10.5)/\sqrt{5} = 4.5/2.24 = 2.008$. The standard normal table reports that Z at 2 and 0.001 = 0.9778; and $1 - 0.9778 = 0.0222$, which is the probability of observing 15 pairs of scores with a positive difference.

The exact one-sided p-value can be calculated using *combinatorics* (combinations of numbers) to assess the probability of observing exactly 15 pairs with positive difference scores. A full explanation of combinatorics is beyond the scope of this chapter. However, for those who may be interested, the procedure to calculate the exact one-sided p-value for the data in Table 6.3 is displayed in Figure 6.17.

$$Z = M - \frac{(n/2)}{\sqrt{n/4}}$$

where Z is the Z statistic for the standard normal distribution,

M = number of pairs with positive differences out of the total number of pairs,

n = total number of pairs,

$\sqrt{(n/4)}$ = square root of total number of pairs divided by 4, where 4 is a constant

FIGURE 6.16
Formula for one-sided p-value for two related samples sign test.

TABLE 6.3

Difference in Time to Pain Relief for Two Headache Relief
Therapies

Case	Therapy 1	Therapy 2	Difference in Hours	Sign
1	30	36	+6	+
2	5	37	+32	+
3	6	38	+32	+
4	15	33	+18	+
5	6	29	+23	+
6	31	4	−27	−
7	45	52	+7	+
8	38	6	−32	−
9	12	60	+48	+
10	7	44	+37	+
11	33	4	−29	−
12	5	32	+27	+
13	3	41	+38	+
14	46	5	−41	−
15	44	45	+1	+
16	18	10	−8	−
17	35	40	+5	+
18	17	42	+25	+
19	4	27	+23	+
20	5	31	+26	+

$$C_{20,15} \times (1/2)^{20} + C_{20,20} \times (1/2)^{20} = 20 \times (1/1049) + 1 \times (1/1049)$$

$$= 0.019 + 0.001 = 0.02$$

FIGURE 6.17
Calculation of exact one-sided p-value for two related samples sign test, where $C_{20,15}$ is read as
"20 choose 15."

Example 6.2 illustrates application of the sign test for related samples using
matched pairs. This example uses the data in Table 6.3 that show the time it
takes for individuals to feel some relief from headache pain under two different therapy methods.

Example 6.2

A nurse practitioner wants to investigate whether patients get faster
relief from headaches with Therapy 1 or Therapy 2. She engages a
group of 40 patients with similar headache problems. The 40 patients
are matched to form 20 pairs based on the average durations of their
headaches. For example, two patients with the average longest duration

time of a headache are matched to form the first pair. The second pair is formed based on the two individuals with the next longest duration time for a headache. Matches of individuals on headache duration time continue to be made until the last match pairs individuals who have the shortest headache duration times.

The clinician then exposes each individual in each pair to Therapy 1 or Therapy 2. Each participant is asked to record the average number of minutes until relief from headaches within a 48-hour period. The null hypothesis is that the median difference between positive and negative responses to Therapy 1 and Therapy 2 is zero. Table 6.3 shows 15 positive difference scores (denoted with plus signs) and 5 negative difference scores (denoted with minus signs).

The SPSS path to access the sign test for related samples is through the Related Samples menu as shown for the McNemar change test in Section 6.1. Figures 6.4 through 6.7 display the relevant dialog boxes for related samples. The path is as follows.

Click Analyze – Click Nonparametric Tests – Click Related Samples – Click the Objective tab – Select Automatically compare observed data to hypothesized in the What is your objective? Panel – Click the Fields tab – Move therapy1 and therapy2 to the Test Fields box – Click the Settings tab – Select Customize tests – Select Sign test (2 samples) in the Compare Median Difference to Hypothesized panel – Click Run

The SPSS results provide a statement of the null hypothesis that was tested, the name of the test conducted, the p-value for the results, and the decision about the null hypothesis. As shown in Figure 6.18, the clinician should reject the null hypothesis of no difference between the medians for Therapy 1 and Therapy 2. The p-value is 0.041 for the two-tailed test. Generally, the two-tailed test is not interpreted.

The one-tailed test result may be obtained by dividing the two-tailed test by 2. A researcher's report of the analysis should include a statement of the null hypothesis, the name of the test, the probability level, and the decision

Hypothesis Test Summary

	Null Hypothesis	Test	Sig.	Decision
1	The median of difference between therapy 1 and therapy 2 equal 0.	Related-Samples Sign Test	0.041[a]	Reject the null hypothesis.

Asymptotic significances are displayed. The significance level is 0.05.

[a] Exact significance is displayed for this test.

FIGURE 6.18
Two-sided p-value is statistically significant for median differences.

to reject or fail to reject the null hypothesis. A summary of the results for Example 6.2 is shown in Figure 6.18.

One can double click on the output to produce a bar graph that shows the positive and negative differences for the related samples sign test. The bar graph in Figure 6.19 displays the differences for Therapy 1 and Therapy 2 for the data in Table 6.3.

Optionally, the Two Related-Samples Sign test may be conducted through the following SPSS path.

> Click Analyze – Click Nonparametric Tests – Click Legacy Dialogs – Click 2 Related Samples – Select therapy1 and move it to the Variable 1 box in the Test Pairs panel – Move therapy2 to the Variable 2 box in the Test Pairs panel – Unclick Wilcoxon if it is checked (Wilcoxon is the default) – Click Sign test – Click Exact – Select Exact – Click Continue – OK

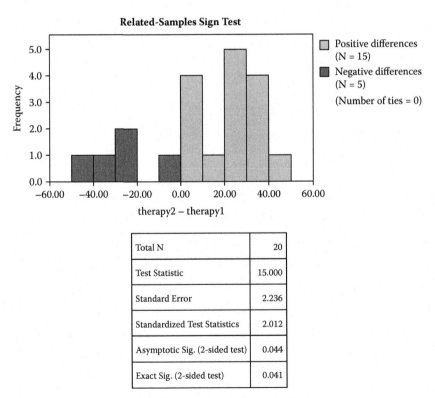

The exact *p*-value is computed based on the binomial distribution because there are 25 or fewer cases.

FIGURE 6.19
Bar graph for sign test differences between Therapy 1 and Therapy 2.

Figures 6.11 through 6.13 are displayed for the McNemar change test in the previous section. These figures display the SPSS dialog boxes to access the Legacy Dialogs path, select the Sign test, and select the Exact test. To conserve space, the figures are not repeated here.

Based on the data in Table 6.3, results of the sign test show that the median time to experience headache relief is less with Therapy 1 than with Therapy 2. These results are reported in Figure 6.20 and show 15 difference scores with plus signs and 5 scores with minus signs. Fifteen or 75% (15/20) of the respondents agreed that Therapy 1 was the most effective in relieving their headaches, whereas 25% reported faster relief with Therapy 2. The test statistic shown in Figure 6.21 indicates that the median difference is statistically significant ($p = 0.021$).

Frequencies

		N
	Negative Differences[a]	5
therapy2 – therapy1	Positive Differences[b]	15
	Ties[c]	0
	Total	20

[a] therapy2 < therapy1

[b] therapy2 > therapy1

[c] therapy2 = therapy1

FIGURE 6.20
Frequency of negative and positive differences in time to headache relief.

Test Statistics[a]

	therapy2 – therapy1
Exact Sig. (2-tailed)	0.041[b]
Exact Sig. (1-tailed)	0.021
Point Probability	0.015

[a] Sign Test

[b] Binomial distribution used.

FIGURE 6.21
One-sided test statistic is statistically significant.

The one-tailed tests are reported for the null hypothesis that positive responses are more probable than negative responses (upper-tailed test). The lower-tailed test is interpreted as negative responses are more probable than positive responses. The null hypothesis must be rejected, and the clinician can conclude that the median time to relief from headache pain is less with Therapy 1 than with Therapy 2. Note that the *p*-value (0.041) for the two-tailed test evaluates the null hypothesis that the probability of the number of positive responses (+) is equal to the number of negative responses (–). Generally, the two-tailed test is not interpreted since researchers are interested in the direction of change rather than whether a change exists.

Box plots visually demonstrate the difference between the two median times. The path to create plots is as follows:

> Click Graphs – Legacy Dialogs – Scroll to Boxplot – Select Simple – Select Summaries of separate variables in the Boxplot panel – Click Define – Move variables (therapy1 and therapy2) to the Boxes Represent box – Click OK

Figure 6.22 shows the response distributions for both therapies. It is obvious from the box plots that the median for Therapy 1 is less than that for Therapy 2. The box plots show that the median time is less than 20 for Therapy 1 and exceeds 30 for Therapy 2.

The sign test for related samples is an alternative to the dependent samples t-test when data are collected on an ordinal scale or the sample size is small. The probability of a difference between groups is based on the number of positive or negative differences between scores for the paired samples. The null hypothesis tests the probability that the frequency of positive differences

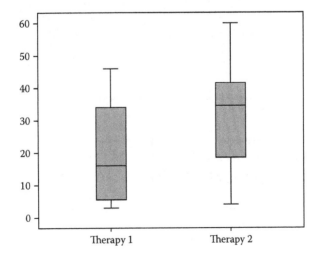

FIGURE 6.22
Distribution of responses for Therapy 1 and Therapy 2.

between pairs of scores is equal to the frequency of negative differences between pairs of scores. Only the direction of change from positive to negative and vice versa is of interest. On the other hand, the Wilcoxon signed rank test is concerned not only with the signs of the differences (+ or –), but also with the sizes of the differences. The Wilcoxon signed rank test is discussed in Section 6.3.

6.3 Wilcoxon Signed Rank Test

The Wilcoxon signed rank (W) test shows the positive or negative direction of the difference between two scores and the relative magnitude of the ranked difference between the scores. Larger differences between scores are given more weight in the rankings than smaller differences. This test is more powerful than the sign test because researchers can discern the sign of the difference between paired scores and also rank the difference scores for a sample.

The test assumes that the medians are equal for two samples in which the paired cases were selected randomly. Samples should be from the same continuous distributions but the test makes no assumptions about the shape of the distribution. Data must be measured on at least an ordinal scale, and the magnitude of the difference scores must be scaled so that scores can be rank ordered.

Like its parametric counterpart, the paired samples t-test, the Wilcoxon signed rank test uses the difference value for each matched pair at two different times or under two different treatments or conditions denoted by X and Y. Each difference (d_i) score may be written as $d_i = X_i - Y_i$. Difference scores are ranked in order of their absolute values. The rank of 1 is assigned to the absolute value of the smallest d_i's, the rank of 2 to the next smallest, etc., until all d_i's are ranked. Recall that the absolute value of a score or number is the number itself without regard for the sign of the number. For example, –5 is given a higher rank than –4 because 4 is smaller than 5. After the scores are ranked, the original sign associated with each difference score is affixed to the difference score that gave it that rank. In other words, the associated sign for a difference score is used to identify ranks that resulted from positive d_i's and ranks that resulted from negative d_i's.

Some cases may result in difference scores that are equal to zero. That is, there is no difference between paired scores for measures taken at two different times on the treatments or conditions. The scores taken at two different times or for two different treatments or conditions would be the same.

If two or more difference scores are equal, the same rank is assigned to the tied scores. For example, if three pairs of scores had differences (d's) of –1, –1, and +1, the scores would all be assigned the average of the rank they would have received if they differed. For the example of –1, –1, and +1, each score would be assigned a rank of 2 because there are three scores. One difference score would have been assigned a rank of 1; one would have been assigned a rank of 2; and the last difference score would have been assigned a rank of 3 [(1 + 2 + 3)/3 = 2]. A rank of 4 would be assigned to the next difference score in the array of scores.

In assigning ranks to tied difference scores, it is important to remember that the average difference score is computed by summing the ranks that would have been assigned had the scores not been tied and then dividing the sum by the number of tied difference scores. Also, it is important to remember that the rank immediately following the average rank for tied scores is the next rank in numerical order. In other words, after a shared rank is assigned, it cannot be assigned again. Another example would be a difference score of 1 for two pairs. If these are the first two difference scores, one would be assigned a rank of 1 and one would be assigned a rank of 2 if these scores were not tied. However, since they are tied, they share a rank [(1 + 2)/2 = 1.5]. Since the ranks of 1 and 2 have been assigned, they cannot be assigned again, so the next largest difference score in the order receives a rank of 3. The W statistic is not affected by assigning the average of ranks to tied scores.

The null hypothesis for the Wilcoxon signed rank test states there is no difference between two treatments or conditions for each pair ($X = Y$). If the null hypothesis is true and there are in fact no differences, one would expect to see some of the larger differences in favor of one variable (X) and some of the larger differences in favor of the other variable (Y). Under a true null, some of the larger ranks result from positive difference values and some result from negative difference values. The distributions of the ranks should be symmetrical such that the differences in the mean and median are zero.

Two statistics are computed to test the null hypothesis. One is for the sum of the ranks for the positive differences (R^+). The other is for the sum of the ranks for the negative differences (R^-). The smaller value of the sum of the ranks is used to compute W.

The following paragraphs explain the procedures to determine the exact *p*-value and the *z*-statistic for the test using hand calculations. These procedures are included for illustration purposes to show how the ranks of the difference scores and the signs of the ranks (+ or –) are used in the analysis. Hand calculations are not recommended. They are tedious, inefficient, and prone to error. However, it is worthwhile to show these calculations to enhance understanding of the process.

The sum of all the ranks of the difference scores is denoted by $n(n + 1)/2$, where n is the number of paired scores. The sum of the negative ranks is $R^- = n(n + 1)/2 - R^+$. After obtaining the sum of the positive ranks (R^+) and the sum of the negative ranks (R^-), we can calculate the exact one-sided *p*-value using the following procedure.

1. Use the smaller of (a) the sum of the ranks for positive differences or (b) the sum of the ranks for negative differences. For example, if the sum of the ranks for negative differences is less than the sum of the ranks for positive differences, list the ranks that are less than or equal to the sum of the negative ranks.

2. Count the number of ranks in the list and sum them. For example, in a list with a sum of five negative ranks, we list the ranks of (0), (1), (2), (3), (4), (5), (1,2), (1,3), (1,4), and (2,3). There are ten ways that ranks can be organized such that each rank is less than or equal to 5.

3. Calculate the number of possible ways that rank assignments can be made using 2 to the nth power, where n is the total number of paired differences. This means the number of ways that a plus or minus sign can be assigned to the pairs. For example, for eight pairs, plus and minus signs can be assigned 2^8 or 256 ways. The 2 represents the two possibilities of sign assignment or sign exchanges (+ to − or − to +) for each pair. This number will be the value in the denominator to calculate the exact p-value.

4. Divide the number of possible ranks that are less than or equal to the observed sum of negative ranks (five for example) by the total number of possible rank assignments denoted as 2^n.

The sum of all the ranks of the difference scores for 8 pairs is denoted by $8(8 + 1)/2 = 36$, and the sum of the negative ranks is $R^- = [8(8 + 1)/2] - 32.5 = 3.5$. The exact one-sided p-value for the data in Table 6.4 is calculated as follows. Total for $R^- = 3.5$, and the following seven ways that ranks can be less than or equal to 3.5 are (0), (1), (2), (3), (3.5), (1,2), and (1,2.5); $N = 8$ and $2^8 = 256$. The exact one-tailed p-value is $7/256 = 0.027$. This procedure will be demonstrated using the data in Table 6.4.

We must reject the null hypothesis that the pretest and posttest scores are equal in favor of the alternative hypothesis that the posttest scores are greater.

Alternatively, the exact p-value can be determined by using a table of p-values for the Wilcoxon signed rank test. The table allows a comparison of the observed value of R^+ to the critical value of R^+ given in the table. The table gives the probability that R^+ is greater than or equal to the observed value of R^+ for a given sample size (n) at an a priori alpha level. To find the p-value, enter the table at the n for the sample size and the value of the sum for the observed R^+.

However, using a table is inconvenient because tables of critical values for the Wilcoxon signed rank test are not commonly included in statistics books. The calculations can become unwieldy and prone to error. For example, listing all ranks that are less than or equal to 9, 10, or 11 negative ranks would certainly invite error! In addition, the tables do not include probabilities for sample sizes greater than 15.

TABLE 6.4

Student Scores on Conflict Resolution

Student	Pretest Score	Posttest Score	Difference Score	Rank of Absolute Difference	Signed Rank of Difference	Sum of Positive Ranks	Sum of Negative Ranks
1	42	45	3	4	+4	4	—
2	29	28	−1	1	−1	—	−1
3	30	41	11	8	+8	8	—
4	28	35	7	6	+6	6	—
5	20	26	6	5	+5	5	—
6	34	32	−2	2.5	−2.5	—	−2.5
7	18	27	9	7	+7	7	—
8	39	41	2	2.5	+2.5	2.5	—
Total				$R^T = 36$	32.5	$R^+ = 32.5$	$R^- = 3.5$

Sample sizes greater than 15 are approximately normally distributed, making it possible to use the normal distribution of z values to determine whether an observed R^+ or R^- is statistically significant. The procedure for calculating the p-value based on a standard normal distribution is more straightforward than the procedure for calculating the exact p-value. The formula for calculating the z-statistic and the solution for the data in Table 6.4 are displayed in Figure 6.23.

According to a table for the standard normal distribution, a z statistic of 2.03 corresponds to a two-tailed p-value of 0.042 at the 0.05 alpha level; the one-tailed p-value is 0.042/2 = 0.021. The decision is to reject the null hypothesis that the ranks are evenly distributed for the two samples.

The previous procedures for determining the statistical significance between two related samples were demonstrations of applications of the formulas. Example 6.3 is based on the data in Table 6.4 that were used to demonstrate hand calculated p-values. The same data will be used to demonstrate the SPSS procedures for obtaining the exact p-value and the z statistic. The next example presents a pretest–posttest design for a research problem.

Example 6.3

A counselor wants to improve the conflict resolution knowledge of her students. She develops a workshop in which she will cover topics such as causes of conflict, strategies for minimizing the causes, negotiation strategies, team building, and improving interpersonal relationships. The counselor believes that the first step to successful conflict resolution is knowledge. She administers a pretest and posttest to eight students selected with the assistance of teachers.

Possible scores for the test range from 0 to 50, with 50 indicating a perfect score and the highest level of knowledge about conflict resolution. Each student is administered the pretest on conflict resolution and

$$Z = R - \frac{n(n+1)/4}{\sqrt{n(n+1)(2n+1)/24}}$$

where R = sum of ranks for negative or positive differences,

n = sample size, and 4 and 24 are both constants

Applying the formula to data in Table 6.4, $Z = -2.03$:

$$Z = 3.5 - \frac{8(9)/4}{\sqrt{8(8+1)(16+1)/24}} = -14.5/7.14 = -2.03,$$

$$Z = -2.03 = -0.4788; 0.50 - 0.4788 = 0.021$$

FIGURE 6.23
Formula for calculating Z statistic for Wilcoxon signed rank test.

then participates in the workshop. At the conclusion of the workshop, the counselor administers the posttest to the students. The steps for the Wilcoxon signed rank test in SPSS are as follows.

Click Analyze – Nonparametric Tests – Legacy Dialogs – Select 2 Related Samples – Highlight pretest and move it to the Variable1 box – Highlight posttest and move it to the Variable2 box in the Test Pairs box – (Optionally, you could click on both variables to highlight them, and hold down the Control Key to move both variables over at the same time.) Click on Wilcoxon test – Click Exact – Select Exact – Click Continue – Click OK

Results for the Wilcoxon signed rank test are shown in Figures 6.24 and 6.25. Figure 6.24 shows two negative rank differences with a mean rank of 1.75 and six positive rank differences with a mean rank of 5.42. Note that the sum of the ranks is also reported. The posttest scores are greater than the pretest scores. Results reported in Figure 6.25 show that the rank differences between the pretest and posttest scores were statistically significant: $z = -2.103, p = 0.04$.

The median of the pretest scores is 29.5; the median of the posttest scores is 33.5. Based on the z statistic and the corresponding p-value, the counselor should reject the null hypothesis of no difference in the pretest and posttest scores for conflict resolution. The results support the claim that the workshop on conflict resolution was successful. This means that students made significant gains in their knowledge of conflict resolution.

An alternative SPSS path to access the Wilcoxon signed rank test is through the Related Samples menu as shown for the McNemar change and sign tests covered in previous sections. Figures 6.4 through 6.7 display the relevant dialog boxes for related samples. The path is repeated here for convenience.

Ranks

		N	Mean Rank	Sum of Ranks
Posttest – pretest	Negative Ranks	2[a]	1.75	3.50
	Positive Ranks	6[b]	5.42	32.50
	Ties	0[c]		
	Total	8		

[a] posttest < pretest

[b] posttest > pretest

[c] posttest = pretest

FIGURE 6.24
Results for Wilcoxon signed rank test showing negative and positive ranks.

Test Statistics[a]

	Posttest – pretest
Z	−2.033[b]
Asymp. Sig. (2-tailed)	0.042
Exact Sig. (2-tailed)	0.047
Exact Sig. (1-tailed)	0.023
Point Probability	0.008

[a] Wilcoxon signed ranks test

[b] Based on negative ranks

FIGURE 6.25
Results of Wilcoxon signed rank test showing significant differences between ranks.

> Click Analyze – Click Nonparametric Tests – Click Related Samples – Click the Objective tab – Select Automatically compare observed data to hypothesized in the What is your objective? panel – Click the Fields tab – Move pretest and posttest to the Test Fields box – Click the Settings tab – Select Customize tests – Select Wilcoxon matched-pair signed-rank (2 samples) in the Compare Median Difference to Hypothesized panel – Click Run

The procedure produces a table showing results of the null hypothesis test for median differences between the pretest and posttest scores. The table is not included here. The results reveal a statistically significant two-tailed test, $p = 0.042$. The decision is to reject the null hypothesis. Plots of the number of negative and positive differences for the pretest and posttest scores may be produced by double clicking on the output. The plots are displayed in Figure 6.26.

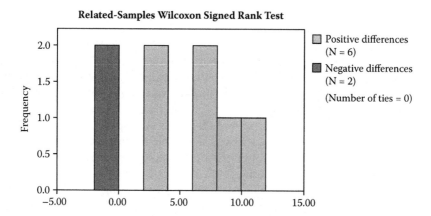

FIGURE 6.26
Plots of pretest and posttest scores for Wilcoxon signed rank test.

The Wilcoxon signed rank test uses the positive and negative ranked differences to show the extent of agreement between related samples. The test is based on the median difference between the paired samples. The median is the central location or midpoint in an array of ranked values. However, more often than not, the true median is located within a range of values called a confidence interval. Section 6.4 discusses the Hodges-Lehmann estimate for the confidence interval for ranked data.

6.4 Hodges-Lehmann Estimate for Confidence Interval

As discussed in Chapter 4, a confidence interval covers a range of values with the lowest value marking the lower limit at one end of the range and the highest value marking the upper limit at the other end. The mean or median difference between paired samples can be computed, along with the probability that the point estimate (mean or median) is located within the interval.

The Hodges-Lehmann confidence interval is useful for the Wilcoxon signed rank test or other tests of related samples that collect data on an interval scale. Hodges-Lehmann confidence intervals can be constructed only on two related samples, both with continuous data. The value of the mean and the median are equal when the distribution is normally distributed, i.e., symmetrical. The upper and lower limits of the Hodges-Lehmann confidence interval are based on the ranked averages of all possible comparisons of paired averages for a data set.

The number of paired estimates is ½ n (n + 1), where n is the number of paired scores. For example, an n of 10 produces 55 estimates [5(11) = 55], which is also the sum of the ranks. An n of 10 would be ranked from 1 to 10, and the sum of the ranks would equal 55 [1 + 2 + 3 + 4 + 5 + 6 + 7 + 8 + 9 + 10 = 55].

The lower and upper limits of the confidence interval can be determined through hand calculations. A table of critical values for the Wilcoxon signed rank test is needed to find the number of paired averages below a value and the number of paired averages above a value that are needed in order to identify the lower and upper limits of the interval, respectively. Such tables are easily accessible in some statistics books and more conveniently on the Internet. The tables list the number of paired averages necessary to reject the null hypothesis at a certain alpha level. Generally, the alpha level is set at 0.05, although 0.10 and 0.01 are also acceptable, depending upon the desired level of conservatism for the width of the interval. The wider the interval, the more likely the point estimate will be included. Specifically, a 1 – 0.01 alpha confidence interval includes 99% of the area under the curve. Surely, in most cases, the point estimate will be in this interval. However, such an interval may not be precise enough to make meaningful decisions about the null hypothesis. The smaller the interval, the more precision (assurance) one can have in identifying the point estimate.

To use a table of critical values, enter the table at n (sample size) and select the appropriate number of paired averages to be excluded based on an a priori alpha level. The lower limit of the interval is the largest value of the lower group of paired averages out of all possible paired averages after the appropriate number of paired averages is excluded. The upper limit of the interval is the lowest value of the upper group of paired averages out of all possible paired averages after the appropriate number of paired averages is excluded. Obviously, without such tables, researchers would not know how many paired averages to exclude before deriving the lower and upper limits. Table 6.5 illustrates how the paired averages are counted to find a 95% confidence interval for the data in Example 6.4.

Example 6.4

Ten individuals enrolled in a university enrichment seminar were asked to complete a pretest on their knowledge of current affairs. After completing the pretest, the seminar facilitator presented brief introductory remarks on a recent documentary film about current affairs and then showed the film to the participants. After viewing the film and before any discussion among the participants, the facilitator asked them to complete a posttest on their knowledge of current affairs. First, the facilitator performed the Wilcoxon signed rank test to determine the extent of the difference between the pretest and posttest scores.

The Wilcoxon signed rank test may be accessed through either of the SPSS paths shown in Section 6.3 of this chapter. SPSS will not allow conduct of the Wilcoxon and the Hodges-Lehmann tests at the same time. It is necessary to conduct each test separately. Figures 6.4 through 6.7

TABLE 6.5

Paired Averages for Hodges-Lehmann Confidence Interval

Lower Limit[a]	Upper Limit[b]
$(31+31)/2 = 31$	$(64+64)/2 = 64$
$(31+35)/2 = 33$	$(61+64)/2 = 62.5$
$(31+37)/2 = 34$	$(61+61/2) = 61$
$(31+42)/2 = 36.5$	$(59+64)/2 = 61.5$
$(35+35)/2 = 35$	$(59+61)/2 = 60$
$(35+37)/2 = 36$	$(59+59)/2 = 59$
$(35+42)/2 = 38.5$	$(56+64)/2 = 60$
$(31+47)/2 = 39$	$(56+61)/2 = 58.5$

[a] Start count with lowest paired average and stop at eighth value.
[b] Start count with highest paired average and stop at eighth value.

display relevant dialog boxes for the Wilcoxon signed rank test. Neither the SPSS paths nor the dialogue boxes are repeated here.

The results of the test shown in Figure 6.27 reveal a statistically significant difference between median scores for the pretests and posttests on knowledge of current affairs (exact one-tailed test = 0.001). Posttest scores were higher than pretest scores for all cases. There were 0 negative ranks and 10 positive ranks. The mean rank was 5.50; the sum of the ranks was 55.

The facilitator was also interested in the confidence interval in which the point estimate (median) may lie. In SPSS, one must enter data for two continuous variables. In other words, entering the difference scores only is not sufficient to calculate the Hodges-Lehmann confidence interval for related samples. The confidence interval for dependent samples may be accessed using the following path.

Click Analyze – Click Nonparametric Tests – Click Related Samples – Click the Objective tab – Select Automatically compare observed data to hypothesized in the What is your objective? panel – Click the Fields tab – Move pretest and posttest to the Test Fields box – Click the Settings tab – Select Customize tests – Select Hodges-Lehmann (2 samples) in the Estimate Confidence Interval panel – Click Run

The results of the Hodges-Lehmann procedure are shown in Figure 6.28. They indicate the point estimate (median value) to be 48, and the lower and upper confidence limits to be 39 and 58, respectively. Repeated sampling from the same or similar populations will produce a point estimate that is included in the confidence interval for 95 out of 100 samples for a 0.05 alpha level. Conversely, there may be a rare occasion (5% of the intervals for an alpha of 0.05) in which the point estimate is not included in the interval.

Ranks

		N	Mean Rank	Sum of Ranks
Posttest – pretest	Negative Ranks	0[a]	0.00	0.00
	Positive Ranks	10[b]	5.50	55.00
	Ties	0[c]		
	Total	10		

[a] posttest < pretest

[b] posttest > pretest

[c] posttest = pretest

Test Statistics[a]

	Posttest - pretest
Z	-2.803[b]
Asymp. Sig. (2-tailed)	0.005
Exact Sig. (2-tailed)	0.002
Exact Sig. (1-tailed)	0.001
Point Probability	0.001

[a] Wilcoxon Signed Ranks Test

[b] Based on negative ranks.

FIGURE 6.27
Results of Wilcoxon signed rank test for knowledge of current affairs.

Confidence Interval Summary

Confidence Interval Type	Parameter	Estimate	95% Confidence Interval	
			Lower	Upper
Related-Samples Hodges-Lehmann Median Difference	Median of the difference between pretest and posttest.	48.000	39.000	58.000

FIGURE 6.28
Hodges-Lehmann confidence interval for Wilcoxon signed rank test in Example 6.4.

Collecting data from a larger sample and calculating the confidence interval again may help resolve this outcome. Reporting confidence intervals is becoming common practice because they provide a range of values within which an estimate lies. Confidence intervals provide more information than only a *p*-value. In fact, many journal editors expect confidence intervals to be included in manuscript submissions.

Chapter Summary

Tests for two dependent or related samples are necessary to ascertain whether one condition or treatment yields higher or lower scores than the other. Subjects are matched or paired on predetermined criteria. Different treatments can be administered to the same individual at different time points or under two different conditions. Sometimes it may be desirable to administer a different treatment or condition to each individual in a matched pair. In both cases, the pretest and posttest scores are compared to ascertain statistically significant differences within the pairs. These tests make no assumptions about the normality of the distributions; however, random selection of pairs is required. The tests presented in this chapter are similar to the parametric dependent samples t-test in which data are continuous in nature.

This chapter discussed three nonparametric tests for dependent samples, the McNemar change test, the sign test, and the Wilcoxon signed rank test. Data for the tests presented in this chapter can be measured on a nominal, ordinal, or interval scale. Each test can be used in a pretest–posttest design. Also, the Hodges-Lehmann estimate for the confidence interval for dependent samples measured on an interval scale was discussed.

The McNemar procedure tests the null hypothesis that the distributions of the before and after responses are equally likely. Results are interpreted as a chi-square test with one degree of freedom. If the calculated value is less than the chi-square table value, one can retain the null hypothesis and assume that the number of changes from positive to negative are not statistically different from the number of changes from negative to positive. If the probability associated with the observed value of chi-square is less than or equal to the predetermined alpha, one must reject the null hypothesis and conclude that the changes are more in one direction than the other.

The sign test for two related samples may be applied in situations in which the researcher's goal is to establish whether two related or dependent observations for treatments or conditions are different. The sign test is inappropriate in situations where unpaired data are used. The null hypothesis states that the median difference between the two related values is equal to zero. In other words, the null hypothesis tests the equality of the two groups.

The Wilcoxon signed rank test considers the extent of the difference between matched pairs, as well as the direction of the difference. It is an extension of the sign test to determine whether the medians or means of related samples differ. Compared to the sign test, the Wilcoxon signed rank test is stronger because more information is used in the calculation of the statistic.

The Hodges-Lehmann 100 $(1 - \alpha)$ percent confidence interval for the median difference between dependent paired samples is based on the average difference between ranked pairs. If the distribution is normally distributed, the mean and median are equal values.

Student Exercises

Data files for these exercises are available online in the file Chapter6data. zip under the Downloads tab at http://www.crcpress.com/product/ isbn/9781466507609. Read each of the following scenarios. Perform the appropriate analyses and respond to the questions.

6.1 Students at a local college organize a seminar to address issues associated with homeless pets. Other students and local residents are invited to participate. Before the seminar begins, all participants are asked whether they would consider adopting a pet from the local humane society. No commitment to adopt is expected or sought. In fact, participants are encouraged not to make immediate decisions about adoption. A one-month minimum wait is recommended so that participants can make thoughtful decisions about adopting a pet. Information such as the kind of pet, expense, time, and care involved should be considered before adoption. Students organizing the seminar are interested only in whether participants change their minds about adopting homeless pets in the interim before and after the seminar. Use the McNemar change test for this problem. This exercise has two data files, one using the raw data and one using the Weighted Cases method.

a. Conduct the appropriate analysis for this study.

b. Write the null hypothesis for this study.

c. Based on the data in the SPSS data file, construct a 2 × 2 table showing the correct frequency in each cell.

d. What proportion of individuals changed their adoption decisions from yes to no?

e. What proportion changed their decisions from no to yes?

f. Provide evidence to support whether the seminar was successful.

g. Write a brief conclusion for this study.

6.2 A biology teacher is interested in students' attitudes toward smoking cigarettes. She administers a 100-point smoking attitudes scale to a group of 16 students in her class. Questions on the scale addressed students' attitudes toward their own smoking and their attitudes toward cigarette smoking in general. For example, one item on the scale was: I would be friends with a cigarette smoker. Another item was: I like the smell of cigarettes. The higher the score, the more favorably students viewed smoking. For the next two class sessions following students' completion of the attitude scale, the teacher presents information and demonstrations about the harmful effects of smoking on health. At the third class meeting, she asks students to complete the smoking attitudes scale again. She is interested in the number and extent to which students changed their attitudes toward smoking from unfavorable to favorable or from favorable to unfavorable. Use the sign test for related samples to perform the analysis.

a. Write the research problem for this analysis using the "lack of information" format.

b. Write the research question(s) for this problem.

c. Write the null hypothesis(es) to be tested.

d. Perform the analysis and apply the decision rule to reject or retain the null hypothesis.

e. Write a brief conclusion for this analysis.

f. Produce a graph of the results.

6.3 A history instructor administers a pretest to his advanced history class to assess students' current knowledge of the Civil War. Fourteen students are enrolled in the course. The teacher will plan the curriculum for the Civil War unit based on the pretest scores. After administering the pretest, he presents the unit of instruction and upon completion of the unit of study, administers a posttest. Both the pretest and the posttest scores are ranked on a scale of 1 to 14, with 1 indicating the highest ranked score in the class, and 14 indicating the lowest. Conduct the dependent samples sign test and respond to the following items.

a. Write the research problem for this study, using the "lack of information" form as presented in Chapter 1.

b. Write the null hypothesis to be tested.

c. What is the design of this study?

d. Report the results of the test and comment on the effectiveness of the unit of instruction based on pretest scores.

e. What was the treatment in this study?

6.4 A restaurant manager asks guests to rate the service they received before and after a major holiday. Only individuals who had been guests in the restaurant one week before and one week after a major holiday were asked to rate the service. They recorded their ratings on a scale of 1 to 10 (10 being the highest rating). Conduct the analysis and respond to the following items.

a. Conduct the sign test for related samples.

b. Was there a statistically significant difference in the number of negative and positive differences before and after the holiday? Give the statistical evidence to support your response.

c. How many of the guests gave the restaurant service a higher rating after the holiday than before the holiday?

    d.   Were there any tied ratings?

    e.   Were the results reported as a binomial distribution or as a normal distribution?

6.5 Students in a middle school were invited to create a poster for a statewide media contest. Thirty-two students developed posters to enter in the competition. Themes for the posters varied from home safety to personal health to pet care. The posters received pre-judge scores on a 100-point scale. The scores were based on the average scores of three judges. The best poster received the highest average score of the three judges. The second best poster received the second highest average of the three judges, and the third best poster received the third highest average of the three judges, etc., until the last poster was scored. Tied scores were allowed in the prejudging. Students received written critiques of their posters. After the prejudging, students had one week to modify their posters and resubmit them in the competition. The posters were scored a second time by a different set of three judges, who also used the same 100-point scale. The second set of judges did not have access to the scores from the first set of judges. The highest average score of the three judges was assigned to the best poster, the second highest average of the three judges was assigned to the second best poster, etc., until all the posters received postjudge scores. Respond to the following questions.

    a.   Write the research questions for this study.

    b.   Write the hypothesis to be tested.

    c.   Conduct the Wilcoxon signed rank test.

    d.   Report and interpret the findings of the analysis.

    e.   Create a graph illustrating the results.

    f.   Report the 95% confidence interval for the test statistic.

6.6 An administrator at a large university recruited 92 faculty members who agreed to participate in a workshop to prepare them to teach courses online. The administrator believes that most faculty members will want to teach online if they learn more about distance education. A workshop on the advantages, disadvantages, logistics, and problems involved in distance education was presented to the faculty. The distance education faculty attitudes scale consisted of 50 items scored on a 100-point scale. Prior to the workshop, faculty attitudes about teaching online were almost evenly split between positive and negative. After completing the workshop, some faculty members changed their attitudes about online teaching from positive to negative, some changed their attitudes from negative to positive, and some did not change their attitudes in either direction.

a. Write the research question(s) for this study.

b. Write the null hypothesis(es) to be tested.

c. Conduct the appropriate statistical procedure to assess whether a change in faculty attitudes was statistically significant.

d. Based on the data in the SPSS data file, construct a 2 × 2 table showing the correct frequency in each cell.

e. What proportion of faculty members changed their attitudes about distance education?

f. Is it possible to produce a confidence interval for the test you conducted? Why or why not?

g. Report the results of the test you conducted. Include a graph with your results.

h. What conclusions can the university administrator reach as a result of your analysis?

6.7 Learning communities are gaining popularity on college campuses. A learning community is organized around a small group of students studying the same course. They share understanding of content with one another, consult with the professor often, and share strategies for solving problems. A mathematics professor wants to test the effectiveness of a learning community for teaching basic concepts in algebra. Nine students are selected for an experiment to assess differences between pretest and posttest scores, using two weeks of participation in a learning community as an intervention. The following partial table of critical values for the Wilcoxon signed rank test should be used with the data for Exercise 6.7 to respond to the items below.

n	Two-Tailed Test (Alpha = 0.05)
5	0
6	0
7	2
8	3
9	5
10	8

a. How many estimates will the nine paired averages produce? (Consult the table.)

b. How many paired samples should be below the lower confidence limit?

c. How many paired samples should be above the upper confidence limit?

    d. Construct a table of paired averages for the 95% Hodges-Lehmann confidence interval.

    e. Use SPSS to produce the Hodges-Lehmann confidence interval.

    f. Interpret the results of the confidence interval.

    g. Was there a statistically significant difference between the pretest and posttest scores?

    h. What can the teacher conclude about learning communities for teaching mathematics?

References

Agresti, A. (2010). *Analysis of Ordinal Categorical Data*, 2nd ed. Hoboken, NJ: John Wiley & Sons.

Albright, B. (2012). The distribution of the sum of signed ranks. *College Mathematics Journal*, 43, 232–236. http://www.jstor.org/stable/10.4169/college.math.j.43.3.232

Cox, D. R. and Donnelly, C. A. (2011). *Principles of Applied Statistics*. New York: Cambridge University Press.

DeGroot, M. H. and Schervish, M. J. (2002). *Probability and Statistics*, 3rd ed. Boston: Addison-Wesley.

Evans, M. J. and Rosenthal, J. (2011). *Probability and Statistics: The Science of Uncertainty*, 2nd ed. New York: W. H. Freeman.

Gibbons, J. D. and Chakraborti, S. (2010). *Nonparametric Statistical Inference*. Boca Raton, FL: Chapman & Hall/CRC.

Hogg, R. V., McKean, J., and Craig, A. T. (2005). *Introduction to Mathematical Statistics*, 7th ed. Upper Saddle River, N J: Prentice Hall.

Hollander, M. and Wolfe, D. A. (1999). *Nonparametric Statistical Methods*, 2nd ed. New York: John Wiley & Sons.

Larocque, D. and Labarre, M. (2004). A conditionally distribution-free multivariate sign test for one-sided alternatives. *Journal of the American Statistical Association*, 99, 499–509. http://www.jstor.org/stable/27590405

Qu, Y. and Easley, K. A. (1997). A method for comparing positive rates of two blood culture systems. *Biometrics*, 53, 1513–1519. http://www.js tor.org/stable/2533518

Rayner, J. C. W. and Best, D. J. (1977). How order affects the sign test. *Biometrics*, 53, 1416–1421. http://www.jstor.or g/stable/2533507

Sirkin, R. M. (2006). *Statistics for the Social Sciences*, 3rd ed. Thousand Oaks, CA: Sage.

Witte, R. S. and Witte, J. S. (2010). *Statistics*, 9th ed. New York: John Wiley & Sons

7

Tests for Multiple Related Samples

Some research studies require that three or more measures be taken over time, under different conditions, or on different sets of individuals or subjects. This chapter presents the Cochran Q test in Section 7.1 and the Friedman test in Section 7.2. Kendall's coefficient of concordance is included as part of both sections and discussed further in Section 7.3. The Cochran Q and Friedman tests analyze differences on repeated measures for a single case or for multiple related samples. In a single case design, every participant is measured on several occasions or under several different conditions. In a matched subjects design, three or more participants are grouped to form a set in which each participant in the set is measured on the same or different variable.

Both the Cochran Q and the Friedman tests are nonparametric analogues to the parametric one-way repeated-measures analysis of variance test (one-way ANOVA repeated measures). These tests are known as blocked comparisons. You will recognize that the Cochran Q test extends the McNemar test to three or more levels of measurement. With only two levels, the Cochran Q and McNemar return the same results. The Friedman test extends the Wilcoxon test to three or more levels of measurement. Kendall's coefficient of concordance W is a descriptive statistic for evaluating concordance or agreement among respondents for measures on multiple objects.

Formulas and explanations are presented to help readers conceptualize test calculations. The SPSS procedures are included to show steps in the conduct of the analysis for each test. Each test is illustrated with an example. The chapter concludes with student exercises to reinforce learning and practice of the tests presented. The data sets for the examples and exercises of this chapter can be retrieved online from the file Chapter7 data.zip under the Downloads tab at http://www.crcpress.com/product/isbn/9781466507609.

7.1 Cochran Q Test

The Cochran Q test is appropriate when dichotomous responses (yes or no, pass or fail, 0 or 1, etc.) are required responses on the dependent measures. Multiple measures may be taken on each individual, such that each individual acts as its own control, or measures may be taken on multiple individuals

$$Q = \frac{(k-1)\, k\Sigma(\Sigma X)^2 - (\Sigma X_T)^2}{k\Sigma X_R - \Sigma X_R^2}$$

FIGURE 7.1
Formula for Cochran's Q test, where k = number of treatments or conditions, $(\Sigma X)^2$ = sum of treatments or conditions on each column squared, $\Sigma(\Sigma X)^2$ = sum of squared treatments or conditions, $\Sigma(\Sigma X_R)$ = grand sum of totals of sums of treatment rows for each case or set, and $\Sigma(\Sigma X^2_R)$ = grand total of squared sums of each row.

in a set. Each individual or set is considered a block and each measure may be considered a treatment or condition. Each block is independent of any other block; however, the measures are dependent within each block.

A matched-subjects randomized block design requires that one of several measures be taken on each individual in a set. The set is then considered to be the block and measures are taken on the treatments or conditions. A set is comprised of individuals matched on some criterion such as test score, relationship, opinion on a specific topic, or some other meaningful variable that is not of interest to the researcher but could influence the outcome of the study. Whether the design calls for multiple measures on each individual or multiple measures for a matched set of individuals, the objective is to test differences across measures.

The Cochran Q test is based on the ranks of the binary responses. It is helpful to visualize this process by recording data in a table with the blocks (cases) in the rows and the treatments or conditions in the columns. The formula and explanations of the component parts in Figure 7.1 will help demonstrate how the data are manipulated to derive the Cochran Q statistic. The test evaluates differences in proportions. The null hypothesis is stated as:

H_0: There is no statistically significant difference in the population proportions.

The alternative hypothesis is stated as:

H_A: The population proportions are different.

Application of the Cochran Q test can be illustrated by the following example.

Example 7.1

A clinical nurse coordinator is interested in students' feedback on the types of experiences they have in three different clinical settings. She wants to test whether the proportions of student ratings in each of the three internship settings are equal. In addition, she wishes to assess the agreement among the student responses.

She recruits 30 students to form 10 sets of students or 3 students per set. Students are placed in sets based on their nursing school admissions test scores. The three lowest scoring students form one set; the next three lowest scoring students form a second set; the next three students form a third set, etc., until the last three highest scoring students are grouped into the tenth set. Each participant within a group is placed within one of three clinical settings: (1) private clinic, (2) hospital, or (3) urgent care facility.

All participants were asked to rate their overall learning experiences in each setting as a 0 to indicate not the best experience to a 1 to denote an excellent experience. For organizational purposes, the clinical coordinator records the data in a table with ten rows—one row for each set of three matched participant—and three columns—one column for each rating of learning experience. The SPSS path for the Cochran Q test is as follows:

Analyze – Nonparametric Tests – Legacy Dialogs – K Related Samples – Unclick Friedman – Click Cochran Q – Hold down the Control key; highlight doctoroffice, hospital, and urgent care; and move them to the Test Variables box – Click Kendall's coefficient of concordance W – Click Exact – Select Exact – Click OK

The SPSS dialog box for the Friedman, Cochran Q, and Kendall's coefficient of concordance W tests is displayed in Figure 7.2. Results of the analysis are reported in Figures 7.3 and 7.4. Figure 7.3 shows that 2 or approximately 20% of the students reported a doctor's office as a best learning experience, 7 (70%) reported a best learning experience at the hospital, and 9 (90%) reported a best learning experience at the urgent care center. Numbers and percents are based on the total number of students (10) assigned to each site.

Figure 7.3 displays statistically significant results for the Cochran Q test $[\chi^2(2, N = 10) = 8.67, p = 0.01]$. We must reject the null hypothesis and conclude

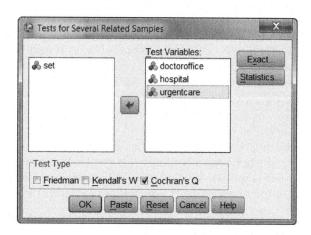

FIGURE 7.2
SPSS dialog box for tests for several related samples.

Frequencies

	Value	
	0	1
Doctoroffice	8	2
Hospital	3	7
Urgentcare	1	9

Test Statistics

N	10
Cochran's Q	8.667[a]
df	2
Asymp. Sig.	0.013
Exact Sig.	0.016
Point Probability	0.013

[a] 0 is treated as a success.

FIGURE 7.3
Results of Cochran's Q test.

that the proportions of responses are different among the students. The exact *p*-value = 0.02. Kendall's coefficient of concordance W (0.43) in Figure 7.4 indicates strong agreement among the student responses. Note that the chi-square value for Kendall's coefficient of concordance W, the chi-square value for Cochran Q's test, and their *p*-values are identical. The advantage of including the Kendall's coefficient of concordance W test is to show the degree of concordance among raters.

Results of the Cochran Q test show that at least one proportion of responses is different from the others; but the results do not show where the difference lies. To determine where the differences are, follow-up pairwise comparisons are necessary. The McNemar test is an appropriate test to follow up a significant Cochran Q test. Recall that the McNemar test is used to assess differences in proportions for dichotomous data. The SPSS path to conducting a McNemar test is included in Chapter 6. Since there are only three comparisons, Fisher's LSD procedure is appropriate to control for Type I error across the three comparisons.

Results of the McNemar follow-up tests indicate that only the comparison between doctor's office and urgent care is statistically significant (p = 0.04). A larger proportion of students (8 or 80%) rated urgent care as a better

Ranks

	Mean Rank
Doctoroffice	1.40
Hospital	2.15
Urgentcare	2.45

Test Statistics

N	10
Kendall's W[a]	0.433
Chi-Square	8.667
df	2
Asymp. Sig.	0.013
Exact Sig.	0.016
Point Probability	0.013

[a] Kendall's Coefficient of Concordance

FIGURE 7.4
Results of Kendall's coefficient of concordance W.

learning experience than a doctor's office. The comparisons between doctor's office and hospital and between hospital and urgent care were not signifi-cant, with p-values = 0.06 and 0.63, respectively

These results may be used to make decisions about placement of future students in internship positions, or at least as a basis for future research into internship placement. The results of the McNemar follow-up tests are shown in Figure 7.5.

7.2 Friedman Analysis of Variance by Ranks Test

The Friedman analysis of variance by ranks test is used to evaluate differ-ences in medians across three or more observations. Multiple measures may be taken on one subject or on sets of subjects such that measures on differ-ent variables or conditions may or may not be rotated among the subjects. The Friedman test is sometimes called a two-way analysis of variance by ranks test with one variable representing the blocks or units (individuals or matched sets) and the other variable representing the observations across

Doctoroffice & Hospital

Doctoroffice	Hospital	
	0	1
0	3	5
1	0	2

Doctoroffice & Urgentcare

Doctoroffice	Urgentcare	
	0	1
0	0	8
1	1	1

Hospital & Urgentcare

Hospital	Urgentcare	
	0	1
0	0	3
1	1	6

Test Statistics[a]

	Doctoroffice & Hospital	Doctoroffice & Urgentcare	Hospital & Urgentcare
N	10	10	10
Exact Sig. (2-tailed)	0.063[b]	0.039[b]	0.625[b]

[a] McNemar Test

[b] Binomial distribution used.

FIGURE 7.5
Results of McNemar follow-up to significant Cochran's Q test.

individuals or sets. Whether measures are taken on individual subjects or subjects are organized into sets, each subject should be independent of any other subject and each set of subjects should be independent of any other set. Random selection is required to assure representativeness of the sample to the population to which one wishes to generalize. Although populations from which the samples were taken do not need to be normally distributed, they should be somewhat symmetrical in that differences between pairwise comparisons are similar with similar continuous distributions.

$$X^2 = \frac{12 \ \Sigma(\Sigma R_T)^2}{n \ k(k+1)} - 3n(k+1)$$

FIGURE 7.6

Formula for Friedman analysis of variance by ranks test, where n = number of blocks (individual cases or sets on rows), k = number of treatments or conditions (number of columns in data set), ΣR_T = sum of ranks for each treatment or condition, and $(\Sigma R_T)^2$ = rank sums squared for each treatment or condition (columns).

These kinds of repeated measures designs are called randomized block designs (RBDs). For single subjects receiving multiple measures, each subject may be considered a block. Likewise, a set of subjects forms a block. Such designs may or may not involve an intervention between measures.

The test is based on data measured on at least an ordinal scale. Data may be recorded in a table with the cases on the rows and the conditions or treatments on the columns. The ranks of measures for each case (row) are ordered with 1 (the smallest rank) to N. Next, the ordered ranks are summed across cases to produce a sum of ranks for each condition or treatment (column). Each summed column is squared and the test statistic can be calculated from this information as a chi-square statistic with treatments minus 1 degree of freedom. The formula to calculate the Friedman analysis of variance by ranks test, with a brief explanation of each component, is presented in Figure 7.6. The formula should give the reader insight into how the test statistic is calculated. The Friedman test evaluates differences in the medians among groups. The null hypothesis is stated as:

H_0: There is no statistically significant difference in the median scores among groups.

The alternative hypothesis is stated as:

H_A: The median scores are different.

Observations should be randomly selected so that they are representative of the population. In addition, scores from one subject or matched pair should be independent of scores from any other subject or matched pair. All data should come from populations with the same continuous distributions, except for perhaps a difference in the medians. Kendall's coefficient of concordance W may also be computed with the Friedman's test to show degree of association. Friedman's test returns a chi-square statistic with degrees of freedom equal to the number of repeated measures minus one. Friedman's two-way analysis of variance test is illustrated by the following example.

Example 7.2

An American history teacher believes that test scores in history can be improved by giving students an opportunity to learn through more than a lecture method. He randomly selects 17 high school sophomores from his lowest achieving history course and designs a study in which students are exposed to three methods of learning about the United States Constitution. Method 1 is simply lecture (lecture) for the entire 50-minute class period. Method 2 consists of a 25-minute lecture followed by a film (lecturefilm) on the Constitution. Method 3 involves only a film followed by small group discussions (filmdiscuss). All lectures address Articles or Amendments to the Constitution. Both films relate to the Constitution.

After each method of instruction, students are given a quiz on the content covered. The three quizzes are of the same level of difficulty and in the same format. Each quiz is scored on a scale of 1 to 20, with 1 representing the lowest score and 20 representing the highest possible score. The teacher records the test scores in a table so that each of the 17 students has three scores, one for each quiz following each method of instruction.

The SPSS steps to performing the Friedman test are the same as those for the Cochran Q test, with two important exceptions: (1) be sure that Friedman is selected and (2) select descriptive statistics, as shown by the following SPSS path.

Analyze – Nonparametric Tests – Legacy Dialogs – K Related Samples – Be sure that Friedman is clicked – Hold down the Control key and highlight lecture, lecturefilm, and filmdiscuss and move these variables to the Test Variables box – Click Kendall's coefficient of concordance W – Click Exact – Select Exact – Click Continue – Click Statistics – Click Quartiles – Click Continue – Click OK

Results of the Friedman test and Kendall's coefficient of concordance are shown in Figures 7.7 and 7.8, respectively. The statistically significant chi-square statistic [$\chi^2(2, N = 17) = 8.46$, $p = 0.02$ (exact p-value = 0.01)] shows that the medians are not from the same distribution and at least one median is different from the others. Therefore, we must reject the null hypothesis of no difference in the medians among the groups. The median for the lecture method is 8.0; the median for the lecture–film method is 11.0; and the median for the film discussion (filmdiscuss) is 12.0. Kendall's coefficient of concordance W (0.25) shows a moderate association among the scores. You will notice that the Friedman test and Kendall's coefficient of concordance W produce identical chi-square values and p-values. The tests are identical except Kendall's coefficient of concordance test returns a measure of agreement among observations.

It is now necessary to conduct follow-up tests to ascertain which medians are different. Pairwise comparisons are conducted using the Wilcoxon test. Fisher's LSD procedure is appropriate to control for Type I error. Refer to Chapter 6 to review the procedures for the Wilcoxon signed ranks test. Results of the Wilcoxon signed ranks follow-up in Figure 7.9 show a significant difference in the median scores for the film and discussion (filmdiscuss) method and the lecture method ($Z = -3.01$, $p < 0.01$). Median scores were higher with the film and discussion method than with the lecture method.

Descriptive Statistics

	N	Percentiles		
		25th	50th (Median)	75th
Lecture	17	5.00	8.00	9.50
Lecturefilm	17	7.00	11.00	12.50
Filmdiscuss	17	9.00	12.00	15.00

Ranks

	Mean Rank
Lecture	1.47
Lecturefilm	2.09
Filmdiscuss	2.44

Test Statistics[a]

N	17
Chi-Square	8.455
df	2
Asymp. Sig.	0.015
Exact Sig.	0.012
Point Probability	0.001

[a] Friedman Test

FIGURE 7.7
Results of Friedman test.

The distribution of the median scores is displayed in the boxplots in Figure 7.10.

Boxplots may be created to exhibit the distribution of median scores. The boxplots displayed in Figure 7.10 may be constructed using the following SPSS path.

> Click Graphs – Legacy Dialogs – Boxplot – Select Simple – Click Summaries of separate variables – Click Define – Hold down the Control key and highlight lecture, lecturefilm, and filmdiscuss – Move these variables to the Boxes Represent box in the Define Simple Boxplot: Summaries of Separate Variables window – Click OK

As displayed in Figure 7.10, the boxplots help us visualize the data by showing the median (horizontal line in each box) and the upper 75th percentile and lower 25th percentile of the distribution (horizontal lines at the tops

Ranks

	Mean Rank
Lecture	1.47
Lecturefilm	2.09
Filmdiscuss	2.44

Test Statistics

N	17
Kendall's W[a]	0.249
Chi-Square	8.455
df	2
Asymp. Sig.	0.015

[a] Kendall's Coefficient of Concordance

FIGURE 7.8
Results of Kendall's coefficient of concordance W.

and bottoms of the plots, respectively). We can see from the boxplots that the most normally distributed plot appears to be that for responses for the film–discuss method, as the median line is close to being equidistant from the upper and lower edges of the boxplot and the whiskers extend about the same distance from the upper and lower boundaries of the plot. If we follow the median lines on each plot to the y axis, it is clear that the medians for the lecture and lecturefilm method are lower than the median for the film-discuss method. Boxplots provide the reader with a quick overview of the distribution of the data.

7.3 Kendall's Coefficient of Concordance (W)

Kendall's coefficient of concordance (W), as illustrated in Sections 7.1 and 7.2, is used to measure the association among multiple observers when more than two objects are rated. The word *object* is used to mean variables that are evaluated or ranked by observers, also called judges. The objects being ranked may be level of achievement or performance, quality of work, attributes of individuals, products, events, or other variables. The objective of the test is to produce an index that indicates the magnitude of the correlation among three or more observers who rank multiple subjects on multiple

Ranks

		N	Mean Rank	Sum of Ranks
lecturefilm – lecture	Negative Ranks	5[a]	7.10	35.50
	Positive Ranks	11[b]	9.14	100.50
	Ties	1[c]		
	Total	17		
filmdiscuss – lecture	Negative Ranks	2[d]	5.00	10.00
	Positive Ranks	14[e]	9.00	126.00
	Ties	1[f]		
	Total	17		
filmdiscuss – lecturefilm	Negative Ranks	7[g]	6.93	48.50
	Positive Ranks	10[h]	10.45	104.50
	Ties	0[i]		
	Total	17		

[a] lecturefilm < lecture

[b] lecturefilm > lecture

[c] lecturefilm = lecture

[d] filmdiscuss < lecture

[e] filmdiscuss > lecture

[f] filmdiscuss = lecture

[g] filmdiscuss < lecturefilm

[h] filmdiscuss > lecturefilm

[i] filmdiscuss = lecturefilm

Test Statistics[a]

	Lecturefilm – lecture	Filmdiscuss – lecture	Filmdiscuss – lecturefilm
Z	-1.685^{b}	-3.010^{b}	-1.328^{b}
Asymp. Sig. (2-tailed)	0.092	0.003	0.184

[a] Wilcoxon Signed Ranks Test

[b] Based on negative ranks.

FIGURE 7.9
Results of Wilcoxon signed rank follow-up test.

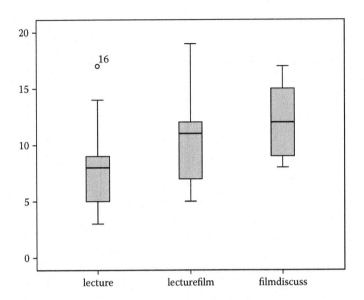

FIGURE 7.10
Median distributions for three teaching methods.

objects (variables). The index of correlation provides a correlation of inter-rater reliability for a set of responses on a set of objects. The several objects to be ranked should be independent of one another and the rankings of the observers should also be independent. Observers are required to express their judgments on all objects by ranking them.

Data should be ordinal in nature. If the data are not ordinal, for example, if the data are in the form of continuous scores, they will be converted to an ordinal scale before Kendall's coefficient of concordance W test is applied. The null hypothesis may be stated as:

H_0: No relationship among the rankings for the set of objects. (One could specify the set of objects, such as test scores, quality of a product, time on tasks, etc.)

Obviously, the alternative hypothesis would state the opposite of the null hypothesis and may be stated as follows:

H_A: A relationship exists among the rankings for the set of objects.

Computation of Kendall's coefficient of concordance W requires that all measures be ranked, summed, and squared for each subject. If there are only a few tied observations, they are assigned the average of the ranks. The deviations of the ranks from the mean rank are computed. This sum of squared deviations is an important component of the formula to calculate Kendall's coefficient of concordance W, as you can see in the formula in Figure 7.11.

$$W = \frac{12S}{m^2 n\left(n^2 - 1\right)}$$

FIGURE 7.11
Formula for Kendall's coefficient of concordance W, where S = sum of squared deviations of ranks from mean rank, m = number of measures, and n = number of objects.

Kendall's coefficient of concordance is very similar to the Spearman rho (ρ) coefficient of correlation statistic and the Friedman test, as both use ordinal data rather than raw scores in their computations. Kendall's coefficient of concordance can be computed directly from the Friedman χ^2 value [$W = \chi^2/m(n-1)$], where m is the number of observers and n is the number of objects being ranked.

The Kendall's coefficient of concordance can also be computed from the Spearman coefficient of correlation, as $W = (n-1)\,\bar{r}_s + 1/n$, where n is the number of observers and $\bar{r}_s$ denotes the mean ranks of the Spearman coefficient of correlation when there are no tied ranks.

The value of Kendall's coefficient of concordance ranges from 0 to +1. The coefficient can never be a negative value since there could not be negative disagreement among three or more observers. Interpretation of the statistic is based on the value of Kendall's W and the p-value. The closer W is to 1, the stronger the agreement among observers; likewise, the closer Kendall's W is to 0, the weaker the agreement among observers. In general, Kendall's W can be interpreted as a correlation with weak, moderate, and strong agreement denoted as 0.10, 0.30, and 0.50, respectively. If Kendall's coefficient of concordance is statistically significant and the agreement is moderate to strong, the Cochran Q test may be conducted to determine where differences lie.

Chapter Summary

The Cochran Q test evaluates differences among multiple distributions when the response variable is dichotomous. The test returns a chi-square value with degrees of freedom equal to number of treatments or conditions minus one. If the chi-square value is statistically significant, usually at an alpha level of 0.05, the null hypothesis should be rejected and multiple comparison follow-up tests are necessary.

The McNemar procedure is used to conduct the follow-up tests. The Friedman test evaluates differences in the median scores for repeated measures on one individual or for sets of individuals. Consequently, it is important that data be on an ordinal or higher scale. If the overall test is significant, then post hoc follow-up tests are necessary. Follow-up tests may be conducted with the Wilcoxon signed ranks test.

Obviously, if the chi-square value is not statistically significant, follow-up tests should not be conducted. Kendall's coefficient of concordance W returns the same chi-square and p-value for both the Cochran Q and Friedman tests. Kendall's coefficient of concordance is a descriptive measure of agreement, rather than an omnibus test. Its purpose is to return the levels of agreement among respondents or observers on the multiple objects being rated, not to identify where any differences lie. Cochran's test may be used to test the hypothesis of differences in proportions among objects.

Student Exercises

Data files for these exercises can be accessed online from the file Chapter7data.zip under the Downloads tab at http://www.crcpress.com/product/isbn/9781466507609. Read each of the following scenarios. Perform the appropriate analyses and respond to the questions.

7.1 A teacher believes that teaching children prediction in story telling will help them develop problem-solving skills. She randomly selects 19 children in the third grade. All students participate in three 20-minute read-aloud sessions with the teacher followed by one of three scenarios: (i) no discussion or question-and-answer period following the reading (scenario1), (ii) a discussion of expectations for each character in the story (scenario2), and (iii) a discussion of expectations for each character in the story and of the plot (scenario3). At the end of each read-aloud session, students are given a brief questionnaire to test their reasoning and accuracy in prediction. The prediction tests are scored from 1 (very low prediction score) to 10 (very high prediction score).

   a. Write the research question and the null hypothesis for this study.

   b. Perform the appropriate statistical procedure to answer the research question and test the null hypothesis.

   c. What are the results of the test?

   d. Are follow-up tests necessary? If so, perform these tests.

   e. Write a results section for this study. Include a graph showing the distributions.

7.2 A wellness counselor randomly selected 21 chronically ill patients from a pool of 45 who agreed to participate in her study of patient well-being. She randomly assigned each patient to one of three wellness techniques to study the effects of the techniques on patients' sense of well-being. The three wellness techniques were Treatment a (trtmenta), 15 minutes in a healing garden; Treatment b (trtmentb), 15 minutes with a counselor; or Treatment c (trtmentc), 15 minutes of individual meditation in the patient's room. The counselor divided the patients into seven groups such that each of the three patients in a group was randomly assigned one of the three treatments. A sense of well-being questionnaire was administered to each patient at the end of each treatment. The questionnaire values ranged from 1 (very insecure sense of well-being) to 25 (very secure sense of well-being).

a. Write the research question and the null hypothesis for this study.

b. Perform the appropriate statistical procedure to answer the research question and test the null hypothesis.

c. What are the results of the test?

d. Are follow-up tests necessary? If so, perform these tests.

e. Write a results section for this study.

7.3 Thirty-five teenagers 13 to 15 years of age were asked their favorite kind of music for relaxation. The choices were rock and roll, classical, country, or hip hop. Each participant was asked to respond to a questionnaire by marking a 0 if they did not prefer a certain kind of music for relaxation or a 1 if they did prefer that kind of music for relaxation.

a. Write the null hypothesis for this study.

b. What is the appropriate statistical procedure for this study? Explain.

c. What are the results of the study? Give the evidence.

d. Perform any follow-up tests that are needed.

e. What can you conclude from this study?

References

Agresti, A. (2010). *Analysis of Ordinal Categorical Data*, 2nd ed. Hoboken, NJ: John Wiley & Sons.

Ary, D., Jacobs, L. C., Razavieh, A., and Sorensen, C. K. (2009). *Introduction to Research in Education*, 8th ed. Belmont, CA: Wadsworth Publishing.

Black, T. R. (2009). *Doing Quantitative Research in the Social Sciences: An Integrated Approach to Research Design, Measurement, and Statistics*. Thousand Oaks, CA: Sage.

Fleiss, J. L. (1965). Queries and notes: a note on Cochran's Q test. *Biometrics*, 21, 1008–1010. http://www.jstor.org/stable/2528261

Friedman, M. (1937). The use of ranks to avoid the assumption of normality implicit in the analysis of variance. *Journal of the American Statistical Association*. 32, 675–701.

Gibbons, J. D. and Chakraborti, S. (2010). *Nonparametric Statistical Inference*. Boca Raton, FL: Chapman & Hall/CRC.

Kashinath, D. P. (1975). Cochran's Q test: exact distribution. *Journal of the American Statistical Association*, 70, 186–189. http://www.jstor.org/stable/2285400

Kvam, P. H. and Vidakovic, B. (2007). *Nonparametric Statistics with Applications to Science and Engineering*. Hoboken, NJ: John Wiley & Sons.

Mack, G. A. and Skillings, J. H. (1980). A Friedman-type rank test for main effects in a two-factor ANOVA. *Journal of the American Statistical Association*, 75, 947–951. http://www.jstor.org/stable/2287186

Runyon, R. P. (1977). *Nonparametric Statistics: A Contemporary Approach*. Menlo Park, CA: Addison-Wesley.

Sprinthall, R. C. (2012). *Basic Statistical Analysis*. Boston: Pearson.

8

Analyses of Single Samples

Researchers do not always have two or more groups to study. Some problems call for analysis of only one group. This chapter presents a discussion of several different tests for analyzing single samples. Section 8.1 discusses the binomial test whose objective is to examine the chance occurrences of outcomes on repeated trials for a binary (dichotomous) variable. Section 8.2 presents the one-sample sign test that is concerned with the direction of the difference between observed values and a specific value. Section 8.3 addresses the one-sample runs test for randomness. The purpose of the runs test for randomness, as the name implies, is to test whether a series of values (a run) is random in a population. Section 8.4 covers the Pearson chi-square test for goodness of fit. The chi-square statistic compares observed frequencies of responses with hypothetical frequencies of responses. Section 8.5 discusses the Kolmogorov-Smirnov one-sample test for examining the difference between an observed sample distribution and an assumed distribution in a population.

Formulas and explanations are presented to help readers conceptualize test calculations. The SPSS procedures are included to show steps in the conduct of the analysis for each test. Each test is illustrated with an example. The chapter concludes with student exercises to reinforce learning and practice of the tests presented. The data sets for the examples and exercises of this chapter can be retrieved online from the file Chapter8data.zip under the Downloads tab at http://www.crcpress.com/product/isbn/9781466507609.

8.1 Binomial Test

The purpose of the binomial test is to ascertain whether outcomes on repeated trials with a binary (dichotomous) variable can be attributed to chance or to a systematic effect. As the name implies, the binomial test is based on the binomial distribution. This is true of small sample sizes. Generally, sample sizes < 20 are considered small. For sample sizes larger than 35, the binomial test is based on the asymptotic p-value for the Z distribution. The binomial distribution has outcomes in one of two groups or classes, but not both. Outcomes are based on a specified number of trials (usually denoted by n or some other letter), and each outcome is the result of one independent trial.

An independent trial is one in which the outcome of one trial does not affect or influence the outcome of any other trial. Only two possible outcomes can result from each trial. These outcomes are called Bernoulli trials. Each outcome is referred to as a success or a failure. The probability that an event will occur is called a success or simply p. The probability that an event will not occur is designated a failure or $1 - p = q$.

A common example of the binomial test is the toss of a fair coin. The only two possible outcomes are heads or tails. Each toss of the coin is independent of any other toss of the coin, so the probability of each outcome is constant (0.5 or 50%). In the coin toss example, we could try to determine the probability of obtaining three heads out of five tosses of a fair coin. The formula in Figure 8.1 is used to find the probability of obtaining a specified number of outcomes (three heads in this case) by a specified number of trials (five independent tosses of a fair coin). Applying the formula to the coin toss example, the result is:

$$\Pr\,(C \geq 3) \sum_{c=1}^{5} \binom{5}{3} = 5^3\,(0.5)^2 = 0.03125$$

Converting 0.03125 to a percentage, the result is calculated as follows:

$$\frac{N!}{C!(N-C)!} = \frac{120}{12} = 10,$$

$$10 \times 0.03125 = 0.3125$$

The resulting probability level of 0.3125 is greater than an alpha level of 0.05, so we may conclude that the outcomes were due to chance. Rarely would one use a formula to calculate probabilities since computers can now perform all of our calculations quickly, easily, and accurately. Instructions for preparing the SPSS spreadsheet to calculate the coin toss example are as follows.

$$\Pr(S \geq C) = \sum_{c=1}^{n} \binom{n}{c} (p)^n (p)^{n-c}$$

where Pr means the probability of obtaining certain outcomes,

S = outcome (count of heads),

C = outcome desired or tested (3),

n = the total number of trials (5),

p^n = the probability of desired outcome out of the total number of trials $(0.5)^3$,

p^{n-c} = the probability of the alternative outcome $(0.5)^2$

FIGURE 8.1
Formula to calculate probability of obtaining three heads out of five coin tosses.

Open the SPSS spreadsheet and type in any name for a variable – (If you do not input a variable name, the default variable name VAR00001 will appear in the Data View) – Open the Data View – Enter any value for the first case of the variable (the value is unimportant). The reason we need to enter a value is that we must use the Transform – Compute function to calculate the probability. (We cannot use this function without at least one value in the Data spread sheet.) Note that 7 has been entered for the variable, score, for the coin toss example.

The SPSS Data View displaying the variable, score, and the value 7 is shown in Figure 8.2. The SPSS path to perform the probability calculation is as follows.

Click Transform – Click Compute Variable – Type prob as the name in the Target Variable box – Click on All in the Function Group – Scroll to Pdf. Binom and Click on the Up arrow to move it to the Numeric Expression box – Replace the question marks in the parentheses with 3, 5, 0.5 (3 for the 3 heads, 5 for the total number of trials, and 0.5 for the binomial distribution for binary variable) – Click OK

The SPSS dialog box for Compute Variable is shown in Figure 8.3. When you open the Data View window, you will see the probability in the column for

FIGURE 8.2
SPSS Data View to calculate probability.

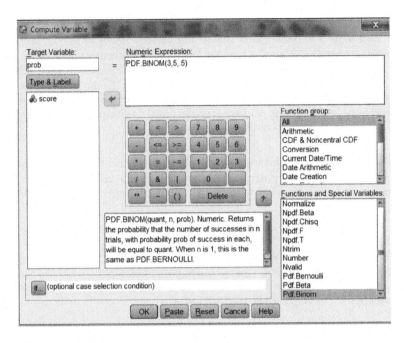

FIGURE 8.3
SPSS dialog box for Compute Variable.

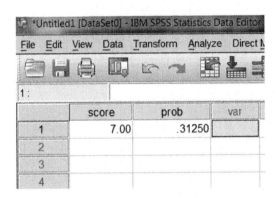

FIGURE 8.4
SPSS Data View showing probability of getting three heads from five tosses of a fair coin.

prob (name entered for the target variable). Increase the number of decimal places in the Variable View to five for the variable prob. The Data View window is displayed in Figure 8.4. The probability of getting three heads among five tosses of a fair coin is 0.31250 or about 31%.

Also, the binomial test can be used to evaluate the probability of obtaining a specified number of independent outcomes or more than the specified number of independent outcomes out of a fixed number of trials (outcomes).

For example, one could ask 10 students whether they preferred an on-campus transit system or on-campus student parking spaces. The prediction may be that students prefer on-campus parking spaces.

What is the probability that we would obtain eight or more preferences for on-campus parking spaces if students were selected randomly? In this case, we want to test the probability of obtaining 8 out of 10 responses plus 10 out of 10 responses in favor of on-campus student parking spaces. This probability is calculated as follows:

$$\Pr = \binom{10}{8} \times 0.5^8 \times 0.5^2 + \binom{10}{10} \times 0.5^{10} \times 0.5^0$$

where $\binom{10}{8}$ means 10 students and 8 responses that we expected (10 choose 8). This is

$$P = (45 \times 0.0039 \times 0.25) + (1 \times 0.00098 \times 1) = 0.04492$$

Note that the calculation of these two independent events is the sum of the probability of both events. Using the SPSS path as in the previous example, we can calculate the sum of the probabilities for two independent outcomes. The SPSS Compute Variable window and appropriate entries for calculating the sum of two independent probabilities are shown in Figure 8.5.

As shown in Figure 8.6, the probability of the two outcomes is 0.04492. Based on the 0.05 level of significance for this test, we can assume that student responses were not due to chance.

Another use of the binomial test is to evaluate probabilities greater than or less than 0.5. For example, what is the probability that a student will guess the correct answer on three of five multiple-choice questions when each question has five possible responses? Applying the binomial formula, we see that the chances of guessing the correct answers on three of five multiple choice questions is five or fewer times out of one hundred. Who said that multiple choice is multiple guess?!

$$\Pr(c = 3) = \sum_{c=3}^{5} \binom{5}{3}(0.2)^3(.8)^2$$

$$= 10 \times 0.008 \times 0.64 = 10 \times 0.0051 = 0.0512$$

Using the SPSS path as in the previous examples, we can calculate the probability. Figure 8.7 shows the Compute Variable dialog box for this kind of problem. Results of the calculation are shown in Figure 8.8. We can use the

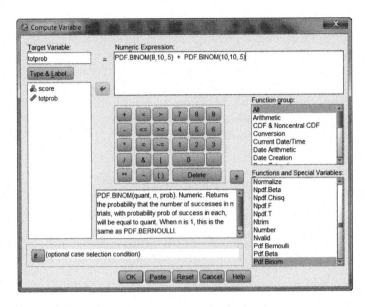

FIGURE 8.5
SPSS Compute Variable dialog box with entries for calculating two independent probabilities.

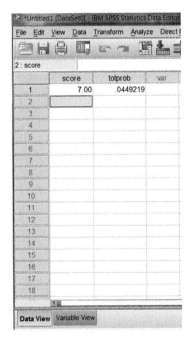

FIGURE 8.6
Probability of two independent outcomes.

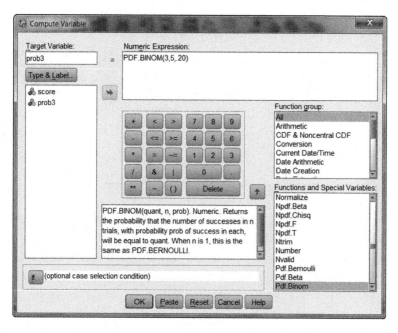

FIGURE 8.7
SPSS dialog box for probability of guessing three correct answers among five items.

FIGURE 8.8
Probability of correctly guessing three out of five multiple choice items.

same procedure as above to calculate the probability of guessing the correct answer of only one of five multiple choice items, each with five response possibilities. This probability is shown in Figure 8.9.

What about the probability of guessing two of five correct responses? Four correct responses? Or correct responses for all five items? The same SPSS procedures as in the earlier examples are used for each of these calculations. Probabilities for correctly guessing one to five correct answers for

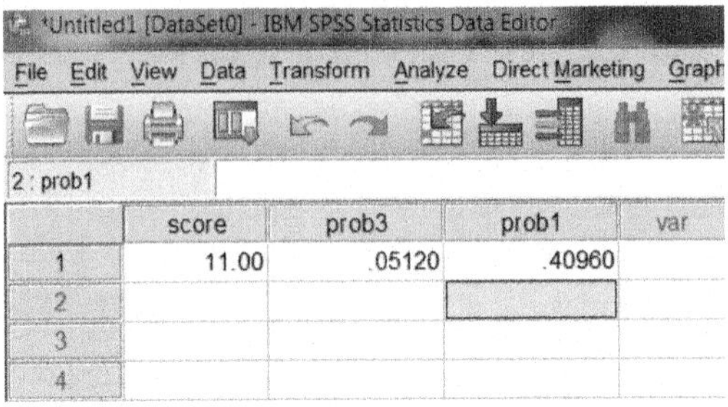

FIGURE 8.9
Probability of correctly guessing one answer (prob1) out of five items.

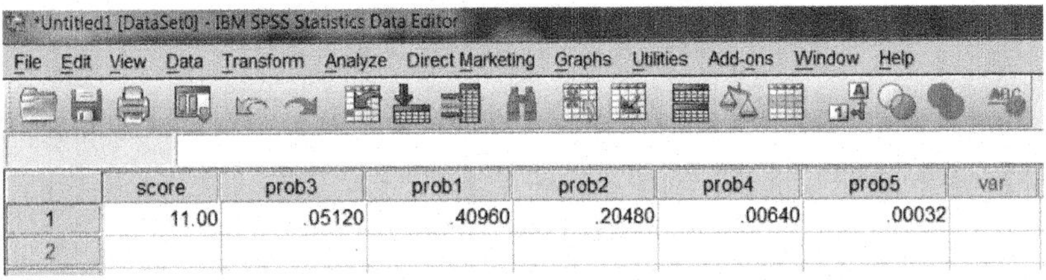

FIGURE 8.10
Probability of guessing correct responses for one to five items, each with five response options.

five multiple choice questions, each with five responses, are displayed in Figure 8.10.

Generally, one does not compute descriptive statistics when the responses take the form of a binary variable. However, the calculations are included here for illustrative purposes.

The mean of a variable with a binomial distribution is given by μ = number of cases or trials × probability ($n \times p$). Using the coin toss example, the mean number of heads that we could obtain in five tosses would be: $\mu = 5 \times 0.5 = 2.5$. Obviously, we cannot obtain 2.5 heads. Using the formula for six tosses ($n \times p$), we could expect the mean number of heads to be 3.

The variance of a variable with a binomial distribution is shown by σ^2 = number of cases or trials (n) × probability of an event occurring (p) × probability of event not occurring (q) or $n \times p \times q$. The variance for our example of five tosses of a fair coin for three heads is calculated as $\sigma^2 = 5 \times 0.5 \times 0.5 = 1.25$. The standard deviation is the square root of the variance. For our example, the standard deviation is: $\sigma = \sqrt{(1.25)} = 1.12$.

8.1.1 Binomial Test and Proportions

Another use of the one-sample binomial test is to assess whether an observed proportion is different from a hypothesized proportion. In this case, the researcher presets a hypothesized proportion. Like other uses of the binomial test, the responses are recorded as dichotomous data in the form of 1s and 0s for pass or fail, success or failure, yes or no, or some other dichotomy.

Example 8.1 illustrates the binomial test used with proportions.

Example 8.1

A teacher may preview a lesson on numbers with 20 students followed by a quiz on memory of the numbers. She may predict that at least one third, one half, three fourths, or some other proportion of students will be successful on the quiz. For this example, the teacher predicts that at least 80% of the students will complete the quiz successfully and sets the cut point at a score of 80 or above to be successful. The null hypothesis for this problem is stated as follows:

H_0: No difference exists in the proportion of students who score above the cut point of 80 on the number test and those who score below the cut point of 80.

The one-sided alternative hypothesis is stated as follows:

H_A: At least 80% of the students will score above the cut point of 80 on the number test.

The SPSS path to calculate different proportions for two groups is as follows.

Click Analyze – Nonparametric Tests – Legacy Dialogs – Binomial – Move quizscore to the Test Variable List box – Click Exact – Select Exact Test – Click Continue – Click Options – Click Descriptive in the Statistics section – Click Continue – Type 80 in the Cut point box in the Define Dichotomy section –Type .80 in the Test Proportion box – Click OK

Figure 8.11 shows the SPSS path for the binomial test, and Figure 8.12 displays the dialog box for the binomial test.

Notice that the default probability changed from 0.50 probability parameter to 0.80 to allow the teacher to specify the probability level at which to compare the two groups (above or below the cut point). Also, note that the Define Dichotomy section of the Binomial Test dialog box allows two choices: allow the dichotomy to be selected from the data or establish a dichotomy by defining a cut point. When scores are in the raw data form, as the quiz scores for this example, it is necessary to define the cut point. If data were entered as 0s and 1s, then defining the dichotomy from the data would be appropriate.

Results of the test are shown in Figure 8.13. Based on the 0.05 alpha level for the exact one-tailed test ($p < 0.01$), we must reject the null hypothesis of no difference in the proportion of students who scored above the cut point of 80 on the number test and those who scored at or below the cut

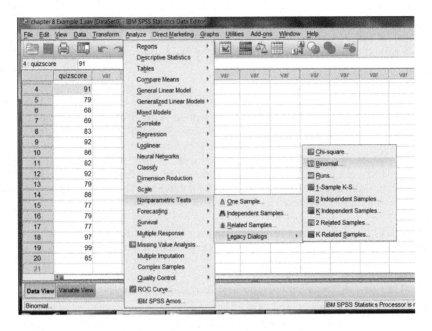

FIGURE 8.11
SPSS path for binomial test.

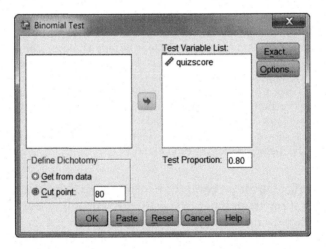

FIGURE 8.12
SPSS dialog box for binomial test.

point. We can claim that a statistically significant difference exists in the proportions between the two groups. Eight students or a proportion of 0.40 scored at or below the cut point, which is significantly different from the hypothesized proportion of 0.20. Twelve students or a proportion of 0.60 is significantly different from the hypothesized proportion of 0.80 who scored

Descriptive Statistics

	N	Mean	Std. Deviation	Minimum	Maximum
Quizscore	20	84.25	8.961	68	99

Binomial Test

		Category	N	Observed Prop.	Test Prop.	Exact Sig. (1-tailed)	Point Probability
	Group 1	≤ 80	8	0.4	0.8	0.000[a]	0.000
Quizscore	Group 2	> 80	12	0.6			
	Total		20	1.0			

[a] Alternative hypothesis states that the proportion of cases in the first group < 0.8.

FIGURE 8.13
Results for binomial test.

above the cut point. The observed proportion of students who scored at or below the cut point is significantly greater than the observed proportion of students who scored above the cut point (0.40 and 0.60, respectively). We can conclude that the predicted success rate (0.80) is significantly different from the observed rate (0.60) at the 0.05 level of significance.

Now that the teacher knows that the predicted success rate of students is different from the observed success rate, she can focus her attention on the causes for the difference and make decisions about the next steps in the instructional process.

As we can see, calculations for the binomial test can be tedious with a hand calculator and have many opportunities for error. Computer software makes such calculations easy and accurate. The binomial test is popular due to its versatility and ease of use. The examples in this chapter represent situations that may be applied to myriad problems in social and behavior sciences. While the examples seem simple, such probabilities can yield important and interesting information about student behavior and achievement.

8.2 One-Sample Sign Test

The sign test, sometimes called the binominal sign test, is the simplest and oldest of the nonparametric tests. It is versatile, easy to use, and involves few assumptions. These characteristics make the test attractive to researchers. The sign test requires only that the data are from a continuous distribution in the population; however, there is no requirement that the data come from any specific shape of distribution. Values must be independent of one another, and each value should be different from other values in the distribution. Tied scores are not used in the analysis.

The sign test can be used for a variety of statistical models. Two of the more popular models are the one-sample sign test and the paired-sample sign test. The one-sample sign test can be used in several situations such as identifying trends in data and testing the null hypothesis of equality of distribution. Use of the sign test for equal distributions is discussed in this chapter. A discussion of the paired-sample sign test is presented in Chapter 6.

The one-sample sign test is a special case of the binomial test and a counterpart to the parametric one-sample t-test. The null hypothesis tests the probability that a random variable from a population that falls above the median of a population is equal to the probability of a random value from that same population falling below the specified value (median). The extent or degree of the difference between the two values is not assessed; only the direction of the difference, greater than or less than a specified value, is used in the analysis.

The one-sample sign test requires only that the researcher mark each response with a plus sign (+) if the response is greater than the population median or a minus sign (−) if the response is less than the population median. Observations that are the same as the median are not considered and are dropped from the analysis. In addition to the median being the measure against which all other values are compared, another value may be used. For example, the hypothesized value could be the mean for the population or a neutral point on a test set by the test developers.

Scores above the median value are assigned a plus sign, and the number of scores to which plus signs are assigned is summed. Scores below the median value are assigned minus signs and the number of minus-sign scores is summed. The total number of positive responses is counted. The letter S is used conventionally to denote the group with positive observations. After the observations are labeled with plus or minus signs, a critical value usually at the 0.05 significance level is used.

A brief explanation of the Wilcoxon signed rank test is included in this section because results for the Wilcoxon test are often desired along with sign test results. The Wilcoxon signed rank test is more powerful than the one-sample sign test because the size of the difference between the median score and the observed score is considered—not just the direction of the difference. However, the Wilcoxon test requires that the distribution be somewhat symmetrical about the median. Like the one-sample sign test, the Wilcoxon test is a nonparametric counterpart to the one-sample t-test. The p-values may differ slightly for the one-sample sign test and the Wilcoxon signed rank test; however, the conclusions generally are the same.

The null hypothesis of equality (probability that a response will be above a specified value is equal to the probability that it will be below the specified value) is tested to ascertain the probability of obtaining a number of observed responses above the median. Likewise, the null hypothesis of equality could be tested to ascertain the probability of obtaining a number of observed responses below the median. The sign test is based on the theory that two outcomes have equal probabilities of occurring ($p = 0.5$). Large sample sizes ($N > 30$) follow a normal distribution. Small sample sizes follow a binomial distribution. For example, if 9 of 12 scores were above a specified cut-off point, the cut-off point could be a neutral point on a standardized scale or some other value set by the researcher. The cut-off point in Example 8.2 is 20. In this case, the neutral point would have been established by the test developers. What is the probability of obtaining 9 of 12 scores above the cut-off point when each trial was constant (50%) and independent? That is, one person's response did not depend on or influence another person's response, and each person had the same opportunity to score above, at, or below the cut-off score. The one-sample sign test is illustrated in Example 8.2.

Example 8.2

A group of 12 individuals reported to have migraine headaches were asked to record the number of minutes it took to relieve their headaches when they took their medications within ten minutes of the onset of headache symptoms. Individuals self-reported the average number of minutes required for the drug to alleviate their headaches over a two-week period. Only participants who experienced two to three migraine headaches per week were eligible for the study. The research question for this study is stated as follows:

> To what extent is the median number of minutes to relieve headache pain equal to 20 when medication is taken within ten minutes of symptoms of the onset of the headache?

The null hypothesis to be tested is stated as follows:

> H_0: The median number of minutes for migraine headaches to be relieved after taking the drug in the first ten minutes of symptoms of the onset of a headache is 20.

A non-directional alternative hypothesis is stated as follows:

> H_A: The median number of minutes for migraine headaches to be relieved after taking the drug in the first ten minutes of symptoms of the onset of a headache is different from 20.

The SPSS path for the one-sample sign test and the Wilcoxon test is accessed through the Two Related-Samples dialog box. The sign test is a single sample and the median value (20 in this case) is treated as a second sample. The SPSS path for the sign test and the Wilcoxon test is as follows:

Click Analyze – Click Nonparametric Tests – Click Legacy Dialogs – Click 2 Related Samples – Move median to the Variable 1 box in the Test

Pairs window and Move minutes to the Variable 2 box – Optionally you can leave the default, Wilcoxon, selected – Click Sign Test – Click Exact Button and select Exact – Click Continue – OK

Figure 8.14 displays the SPSS path for the one-sample sign test. Notice in the Data View window that median is a variable. The values are not visible; however, each case has the same value, i.e., 20. Figure 8.15 shows the dialog box for a two related-samples test in which one can select the sign and Wilcoxon tests. Notice in the figure that median is a variable in the Test Pairs box.

As stated earlier, the results for the Wilcoxon sign test are close to the results for the one-sample sign test. If we left the Wilcoxon option selected, based on the asymptotic p-value for a two-tailed test, the null hypothesis can be rejected at the 0.05 level ($p = 0.01$). We can thus conclude that the median number of minutes until headache relief is different from 20. If we wish to test the hypothesis that the number of minutes to headache relief exceeds 20 (right-sided test), which the mean rank (7.20) supports, we interpret the one-tailed test.

The p-value for the exact one-tailed test is < 0.01, confirming that the median number of minutes to headache relief is greater than the proposed median value of 20. If we wished to assess the p-value for a left-sided test (median number of minutes to headache relief is less than 20), which the

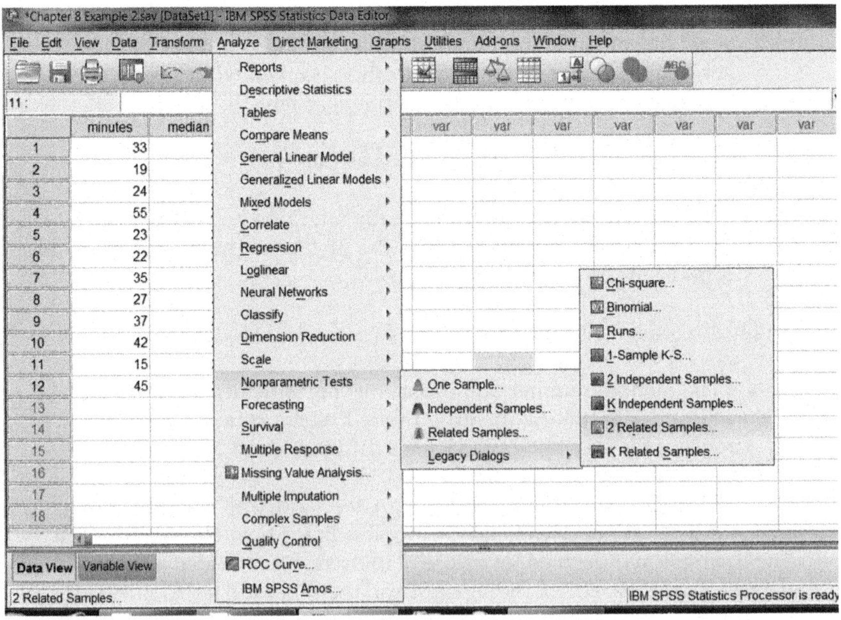

FIGURE 8.14
SPSS path to one-sample sign test.

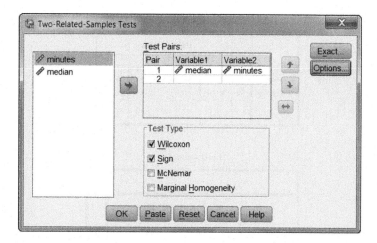

FIGURE 8.15
SPSS dialog box to access sign test and Wilcoxon test.

data do not support (mean rank = 3.00), the calculation is 0.010/2 = 0.005 and 0.005 − 1 = 0.995. Also, notice that the Wilcoxon signed-ranks test produces a Z value (−2.59) based on the negative ranks.

The negative ranks are used because the smaller of the two N's is used to interpret the results. For the group scoring below 20 as the median number of minutes to headache relief, N = 2. N = 10 for the group scoring above 20 minutes. The results of the Wilcoxon test are not shown here. You will see later that the decision to reject the null hypothesis is the same for the one-sample sign test as for the Wilcoxon test.

Figure 8.16 displays the results of the one-sample sign test. Results show 10 positive differences between the number of minutes for headache relief and a median of 20 and two negative differences—the same numbers for both groups as for the Wilcoxon test.

As shown in Figure 8.16, the median number of observed minutes to headache relief (30) is greater than the hypothesized median number of minutes (20). However, to assess the statistical significance of the difference between the positive and negative mean ranks, the test statistics must be interpreted. The exact two-tailed test is statistically significant beyond the 0.05 level ($p = 0.04$), which allows rejection of the null hypothesis and the conclusion that the median number of minutes to headache relief is significantly different from 20. This conclusion is supported by the data.

To obtain the p-value for the one-tailed test, divide 0.039 by 2 to get a one-tailed p-value of 0.019 (shown in Figure 8.16). Results support the alternative hypothesis that the median number of minutes to headache relief exceeds 20. To obtain the p-value for the number of minutes less than the median (which the data do not support, as we have only two negative differences), the formula is the same as for the Wilcoxon signed ranks test: 0.039/2 − 1 = 0.98.

Descriptive Statistics

	N	Percentiles		
		25th	50th (Median)	75th
Median	12	20.00	20.00	20.00
Minutes	12	22.25	30.00	40.75

Frequencies

		N
Minutes-median	Negative Differences[a]	2
	Positive Differences[b]	10
	Ties[c]	0
	Total	12

[a] minutes < median
[b] minutes > median
[c] minutes = median

Test Statistics[a]

	minutes – median
Exact Sig. (2-tailed)	0.039[b]
Exact Sig. (1-tailed)	0.019
Point Probability	0.016

[a] Sign Test
[b] Binomial distribution used.

FIGURE 8.16
Results of one-sample sign test.

The one-sample sign test is based on the number of values above or below a pre-specified value—usually the median. The median value may be based on an empirical or theoretical basis discovered in the research or on a value identified via a standardized test procedure. The versatility, ease of use, and straightforward interpretation make the one-sample sign test a common tool for test developers and other researchers. Another one-sample test that uses dichotomous variables is the one-sample runs test for randomness. This test is appropriate for determining whether a series of values is random. The one-sample runs test for randomness is discussed in Section 8.3.

8.3 One-Sample Runs Test for Randomness

The assumption of randomness is important to parametric procedures. The one-sample runs test for randomness (one-sample runs test) permits researchers to check the assumption of randomness statistically. The one-sample runs test assesses whether a series of values for a dichotomous variable are random. The values may represent an attribute variable such as values that are binary and coded as either a 0 or 1, or the values may be interval data that are divided into two groups. Specifically, the purpose of the runs test is to evaluate whether the order of occurrence of responses for an event or condition is random. The null hypothesis for randomness is stated as follows:

H_0: The sequence is random. (It is recommended that clarity be provided for a sequence such as the sequence of yes responses, the sequence of alternative conditions, etc.)

The alternative null hypothesis is stated as follows:

H_A: The sequence is not random.

For a dichotomous (binary) variable, a run occurs each time the variable value changes in a series as in the following series of binary values: 1111, 0000, 11, 0. Essentially, we count the number of times the series changes from one value to another. In other words, a run is an occurrence of an uninterrupted series of the same value. A single occurrence is counted as a run of one. In this example we have four runs. The first run consists of four 1s; the second run is the series of four 0s; the third run has two 1s; and the fourth run is a single 0. In a series of values where 0 represents one group or value and 1 represents the other group or value, we may observe the values in many different orders. The order in which data are collected is irrelevant; however, it is important that the data be entered into a spreadsheet for analysis in the order in which they were collected. Grouping or prearranging the data generally affects the number of runs and leads to questions about randomness, credibility of the results, and integrity of the researcher.

The variable or attribute to be assessed by the runs test need not be always dichotomous. If the data are interval, a cut point can be specified. The cut point can be the mean, median, or mode of the distribution of values or some other value determined by the researcher. Variables that are not dichotomous are treated as if they are dichotomous based on the cut point. Values equal to or greater than the cut point form one group generally denoted with a value of 1, and values less than the cut point form a second group generally denoted by 0. Example 8.3 illustrates use of the runs test for dichotomous data.

Example 8.3

A family counselor collected data on marital status and assigned 0 to married status and 1 to not-married status. The counselor observed the following series of 0s and 1s for the first 15 clients who sought family counseling after the New Year's holiday: 000, 11, 00, 11, 0, 1, 0, 1, 0, 1. Is this series random?

First the counselor would count the number of runs. The first three 0s count as one run; next, the two 1s count as a second run; the two 0s count as a third run; followed by the two 1s that count as a fourth run; the single 0 counts as the fifth run. The next single 1 serves as the sixth run; the next single 0 takes the total to seven runs; then another single 1 becomes the eighth run; another 0 becomes the ninth run; and the final 1 serves as the tenth run. The counselor observes 10 runs in a sequence of 15 values. With only 15 observations, the counselor may be able to make a judgment about the randomness of this series of values. However, subjective observation may lead others to question the judgments made and the accuracy of the conclusions. The SPSS path for checking randomness of responses for marital status data is as follows.

Click Analyze – Click Nonparametric Tests – Click Legacy Dialogs – Click Runs - Unclick Median – Click Custom in the Cut Point panel and type 1 in the Custom box – Move maritalstatus to the Test Variable List box – Click the Exact button – Click Exact – Click Continue – Click OK

The SPSS path to access the runs test is the same as that for accessing the sign test and the Wilcoxon test shown in Figure 8.14, except that Runs should be selected instead of 2 Related Samples. The SPSS dialog box for the runs test is shown in Figure 8.17.

An alternative path to the runs test produces a hypothesis test summary that states whether the sequence of values is random. The results produce the decision to reject or fail to reject the null hypothesis. The path is included here for illustration purposes:

Click Analyze – Click Nonparametric – Click One Sample – Click the Test sequence for randomness in the One-Sample Nonparametric Tests dialog box (be sure that the Objective tab is active) – Click Run

The dialog box for conducting the runs test using the above path is displayed in Figure 8.18. If you are interested in alternative paths to conduct other one-sample nonparametric tests, click the Settings tab to see the options available.

Results for the runs test indicate that the family counselor should fail to reject the null hypothesis at the 0.01 alpha level, ($Z = 0.56$, $p = 0.58$, and exact p-value = 0.45). The counselor can conclude that there was no pattern in the marital status of clients who sought family counseling in the first days following the New Year. The results of the runs test are included in Figure 8.19.

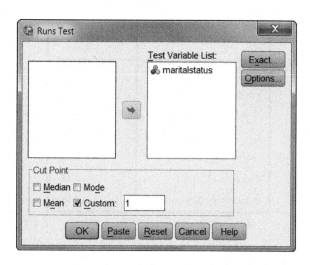

FIGURE 8.17
SPSS dialog box for runs test.

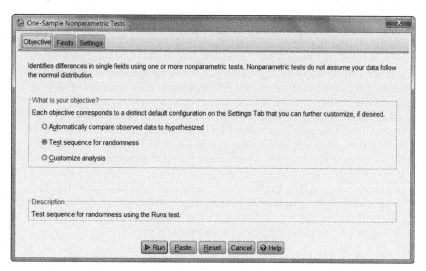

FIGURE 8.18
SPSS dialog box to test sequence for randomness using runs test.

Situations may arise when a researcher wishes to verify the randomness of responses measured on an interval or ratio scale. Responses for participant attitudes, values, opinions, or preferences measured on a Likert-type scale can be checked for randomness. When responses are not dichotomous, a cut point in the response set is used to classify responses into two groups, and the runs test is conducted on the newly created dichotomous variable. The cut point can be used for small or large sample sizes. The following example demonstrates the runs test for interval data.

Runs Test

	Maritalstatus
Test Value[a]	1.00
Total Cases	15
Number of Runs	10
Z	0.556
Asymp. Sig. (2-tailed)	0.578
Exact Sig. (2-tailed)	0.445
Point Probability	0.163

[a] User-specified.

FIGURE 8.19
Results of one-sample runs test for randomness.

Example 8.4

The manager of a gym is offering a special promotion to attract new members and hypothesizes that individuals living in retirement homes may be potential prospects. The manager contacts three local retirement homes and arranges for a group of 25 residents to tour the gym. At the end of the tour, visitors are asked to indicate their likelihood of joining the gym based on several criteria such as current level of exercise, cleanliness of the gym, friendliness of gym personnel, and condition of equipment. The gym impressions Likert-type questionnaire included 20 items. Each item was based on a five-point scale, ranging from strongly agree to strongly disagree. The highest possible score on each item was 5 for strongly agree, and the lowest possible score on each item was 1 for strongly disagree. The highest possible total score was 100 and the lowest possible total score was 20. The manager is interested in whether the visitor responses on the instrument were random or followed a systematic pattern. The null hypothesis is stated as follows:

H_0: Visitor responses on the gym impressions questionnaire are random.

The alternative hypothesis is:

H_A: Visitor responses on the gym impressions questionnaire follow a systematic pattern.

The SPSS path to conduct the one-sample runs test for interval data is as follows:

Click Analyze – Click Nonparametric – Click Legacy Dialogs – Click Runs – Move joingym to the Test Variable List box – Note that Median is already selected in the Cut Point panel of the dialog – Click Exact – Select Exact – Click Continue – Click OK

Review Figure 8.17 for the dialog box for the runs test. Note that the median was unselected for the runs test for dichotomous data. The median is the default and should remain selected as the cut point for interval data.

Results of the runs test for interval data are displayed in Figure 8.20 and indicate that the median (51) was used as the cut point for the 25 scores. This value is listed on the output as the test value used to transform the data into a binary sequence of 0s and 1s. The minimum number of possible runs is two (all scores below 51 appear in a consecutive sequence and all scores equal to 51 and above 51 are clustered together in a second sequence). The maximum number of runs is 25 if the scores below and above the median alternate in the series in a pattern such as 0, 1, 0, 1, 0, etc.

There were 12 runs for these data; the sequence was 0, 1, 1, 1, 0, 1, 1, 1, 1, 1, 1, 0, 0, 0, 1, 0, 0, 0, 1, 0, 0, 1, 0, 0, 1. The probability of obtaining 12 or fewer runs is 0.68 as shown by the exact significance for the two-tailed probability (p-value). The asymptotic p-value is 0.69. The non-significant p-value indicates that we must retain the null hypothesis and conclude that the responses are random. It is not necessary to report both the asymptotic significance level and the exact significance level.

The asymptotic and exact test p-values are not always the same although in this case the difference is negligible. In most cases, the difference between

Runs Test

	Joingym
Test Value[a]	51
Cases < Test Value	12
Cases ≥ Test Value	13
Total Cases	25
Number of Runs	12
Z	−0.401
Asymp. Sig. (2-tailed)	0.688
Exact Sig. (2-tailed)	0.683
Point Probability	0.141

[a] Median

FIGURE 8.20
Results of runs test for interval data.

the asymptotic and exact *p*-values is not large enough to affect the decision about the null hypothesis.

The purpose of the runs test, whether for small or large sample sizes, or dichotomous or interval data, is to assess the probability of a series of responses or scores being random. The chi-square goodness-of-fit test discussed in Section 8.4 is another test for single samples. It evaluates the probability of a difference between observed responses in one category or group and expected responses into the same category or group.

8.4 Pearson Chi-Square Test for Goodness of Fit

The chi-square test for goodness of fit is one of the most common nonparametric procedures. The test evaluates whether there is a statistically significant difference between observed scores or responses for a sample and expected or hypothesized scores or responses in a population. The Pearson chi-square test for goodness of fit is based on the assumption that observed scores fall randomly into one category or another, and the chance of a score falling into a particular category can be estimated. In other words, the one-sample chi-square test may be used to check the extent to which a distribution of observed (sample) scores fits an expected or theoretical distribution. Three other assumptions are important to the test: the data are from a randomly selected sample from the population, responses are independent of one another, and no cell has fewer than five expected frequencies. It is crucial to the credibility of the results that all the assumptions are met.

The one-sample chi-square goodness-of-fit test can be used in several situations. For example, it can be used to check normality of a distribution and also to evaluate the difference between observed data and some other distribution specified in the null hypothesis. In addition, the test can be used to assess differences between two or more sets of observed data. The focus of this section is on the chi-square goodness-of-fit test for agreement between the observed frequency of responses and the expected frequency of responses.

Two or more categories may be tested for goodness of fit. The test is designed to assess the extent of agreement between the observed and expected outcomes in each category. Unless the researcher has information or a rationale to the contrary, the expected frequencies represent an equal proportion of cases in each category. Thus, the expected frequency may be calculated by $E_i = N/k$ where E_i represents the expected frequency, N represents the total number of cases, and k represents the total number of categories. The probability that the observed number of responses could have come from a population with the expected frequencies is

$$\chi^2 = \sum \frac{(f_o - f_e)^2}{f_e}$$

FIGURE 8.21

Formula to calculate chi-square goodness-of-fit test, where f_o is the observed frequency of cases, f_e is the expected frequency of cases, and $(f_o - f_e)^2$ is the squared difference between the observed and expected value in each cell.

tested with the chi-square goodness-of-fit test formula displayed in Figure 8.21.

Expected frequencies are the proportions of responses hypothesized to fall in each of the categories represented in the population. Data for the chi-square goodness-of-fit test can be recorded in a $1 \times C$ contingency table. Each cell contains data that represent a category of response. Each cell (category) has two entries, one for the observed score or response and one for the expected score or response. The chi-square statistic may be derived by summing the squared differences between each observed and expected frequency and dividing by the corresponding expected frequency.

A small difference between the observed and expected frequencies produces a small chi-square statistic. On the other hand, a large difference between the observed and expected frequencies produces a large chi-square statistic; and the less likely it is that the observed frequencies came from the population on which the expected frequencies were based. Example 8.5 illustrates a problem that may be investigated using the one-sample chi-square test.

Example 8.5

A common problem on most university campuses is lack of parking spaces. To address this problem, a major university instituted a campus transit system five years ago to transport students from common parking areas to specified locations on campus. The administration is not totally convinced that the transit system is working efficiently and effectively. On the other hand, the campus manager for the transit system believes that the transit system is a safe, efficient, and economical mode of transportation for students.

The manager for the transit system would like to get the students' perspective on the transit system. If he can show the administration that the system is accomplishing what it is designed to do, he may be able to increase the number of buses and drivers, thereby providing more services for students and more jobs for drivers. With permission of the Institutional Review Board and the university administrator, the manager collects data from the students who use the transit system. He uses 4×6 note cards that ask students to mark one of three responses: I have a positive perception of the campus transit system; I have a negative perception of the campus transit system; or I feel neutral (neither positive nor negative) about the campus transit system. The number of expected frequencies of responses is assumed to be the same for each category of response. The manager makes 150 such cards and distributes 10 each to 15 randomly selected drivers who cover different routes. Drivers are instructed to give the cards to 10 students at random as they depart the buses. The drivers are cautioned against handing cards out to small groups of students riding together to avoid bias in responses. Students are instructed to drop their responses in a closed container inside the building of their destination. Of the 150 cards, 123 (82%) were returned. The manager expects the same number of responses for each of the categories of responses. The null hypothesis for this study is stated as follows:

H$_0$: There are no differences in proportion of responses for each of the categories: (1) positive, (2) negative, or (3) neutral.

The alternative hypothesis is stated as follows:

H$_A$: There are differences in the proportion of responses for each of the categories: (1) positive, (2) negative, or (3) neutral.

Data are coded +1 for positive perspective, 0 for neutral perspective, and –1 for negative perspective. With 123 cases, it is appropriate to use the weight cases method to prepare the data for analysis. The SPSS steps for the method are presented in Chapter 3. The steps are repeated here for convenience. Refer to Figure 3.4 to see the Weight Cases dialog box.

Click Data – Click Weight Cases – The Weight Cases dialog box will open – Select Weight cases by – Move the count variable to the Frequency Variable box – Click OK

After the cases are weighted, use the following SPSS path for the one-sample chi-square goodness-of-fit statistic.

Click Analyze – Click Nonparametric Tests – Click Legacy Dialogs – Select Chi-Square – Move perception to the Test Variable List box – Click Exact – Click Exact in the Exact Tests dialog box – Click Continue – Leave the defaults, Expected Range, Get from data and Expected Values, All categories equal, selected – Click OK

Results of the analysis show statistically significant differences between the observed and hypothesized proportions for each of the categories [$\chi^2(2, N = 123) = 11.17, p < 0.01$, and exact *p*-value < 0.01]. Therefore, the transit manager can conclude that the observed distribution of the proportion of scores in each category is different from the expected proportions.

Observed frequencies of responses are fewer than expected (36 observed compared to 41 expected) for responses marked –1, and observed frequencies of responses are fewer than expected (29 observed compared to 41 expected)

for responses marked 0. Observed frequencies are less than expected for students with negative perceptions of the transit system and students who are neutral. Observed frequencies of responses are greater than expected (58 observed compared to 41 expected) for responses marked +1. The degrees of freedom are derived from the number of levels (cells or categories) minus one; thus, in our example with three categories (negative, neutral, positive), we have two degrees of freedom. Figure 8.22 displays the results for the chi-square test.

With a significant chi-square statistic and more than two response categories, follow-up tests are needed to identify whether the difference in proportions between any two of the groups is due to a disproportionate number in the third group. For example, to test whether there is a statistically significant difference in proportion of students who have negative perceptions of the transit system and those who are neutral, only the negative responses and neutral responses need to be compared. We must select the appropriate cases before conducting the follow-up tests. Use the following path in the SPSS data window to select cases.

Perception

	Observed N	Expected N	Residual
Negative	36	41.0	−5.0
Neutral	29	41.0	−12.0
Positive	58	41.0	17.0
Total	123		

Test Statistics

	Perception
Chi-Square	11.171[a]
df	2
Asymp. Sig.	0.004
Exact Sig.	0.004
Point Probability	0.000

[a] 0 cells (0.0%) have expected frequencies less than 5. The minimum expected cell frequency is 41.0.

FIGURE 8.22
Results of Pearson chi-square test for goodness of fit.

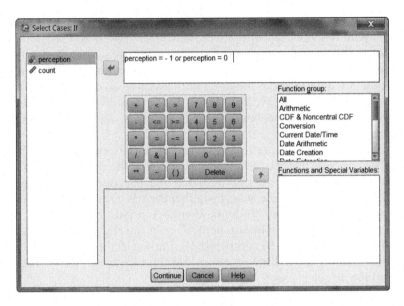

FIGURE 8.23
SPSS Select Cases dialog box with negative and neutral responses selected.

> Click Data – Click Select Cases – Select If condition is satisfied – Click If in the Select Cases dialog box – Highlight perception and move it to the blank box in the Select Cases: If dialog box – Type = –1 or after perception – Move perception to the blank box again and it will appear after the word or and then type = 0 in the Select Cases: If dialog box – Click Continue – Click OK

Figure 8.23 shows the Select Cases: If dialog window with the appropriate perception response categories given for the negative and neutral responses (–1 and 0). After selection of cases for the categories to be compared, the Data View window will appear as it is in Figure 8.24. With the desired response categories selected, the first follow-up test can be conducted.

Chi-square results of the follow-up test to evaluate differences in proportions between the negative and neutral responses were non-significant [$\chi^2(1, N = 65) = 0.75$, $p = 0.39$]. The transit manager can conclude that the proportions of responses for the two categories are not different. Chi-square results were statistically significant for the follow-up test to evaluate differences in proportions between negative and positive responses [$\chi^2 (1, N = 94) = 5.15$, $p = 0.02$, exact p-value = 0.03]. The transit manager can conclude that the proportions of responses for the negative and positive responses were different.

Observed frequencies for negative responses were less than the expected frequencies (36 and 47, respectively), and observed frequencies for positive responses were greater than the expected frequencies (58 and 47, respectively). Chi-square results were statistically significant for the follow-up

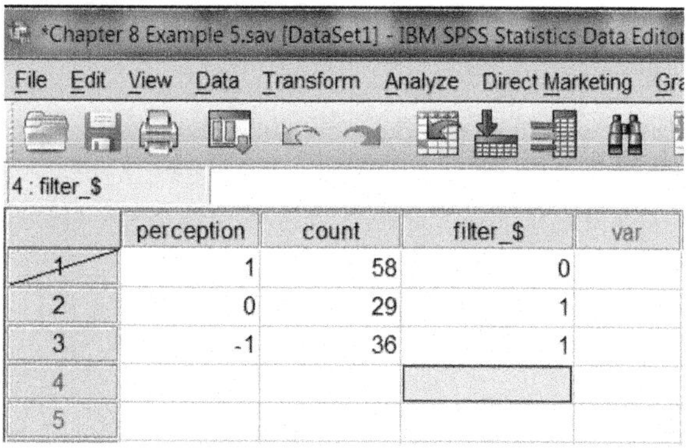

FIGURE 8.24
SPSS Data View with negative and neutral responses selected.

test to evaluate differences in proportions between neutral and positive responses [$\chi^2(1, N = 87) = 9.67, p = < 0.01$]. Observed frequencies for neutral responses were less than the expected frequencies (29 and 43.5, respectively), and observed frequencies for positive responses were greater than the expected frequencies (58 and 43.5, respectively).

The transit manager can conclude that the proportions of responses for the neutral and positive responses were different. The discrepancy between the observed and expected frequencies in favor of students' positive perceptions of the transit system helps support the transportation manager's hypothesis that the transportation system operates efficiently and effectively. Student responses to the transit system survey can be graphically displayed in a bar chart. The SPSS path to create the graph is as follows.

> Click Graphs – Click Legacy Dialogs – Click Bar – Click Simple – Select Summaries for Groups of Variables Data in Chart Area section – Click Define – Move perception to the Category Axis box in the Define Simple Bar: Summaries for Groups of Cases dialog box – Click OK

Figure 8.25 shows the SPSS dialog box to create the bar chart. The bar chart is displayed in Figure 8.26.

In Example 8.5, the number of expected responses was believed to be equal for each of the three levels. If, however, we had an empirical or theoretical basis for hypothesizing that the number of expected frequencies for two levels combined was different from the number of expected frequencies for a third level, we would need to create a new variable to sum the values of the two levels that are expected to be different from the third level. The chi-square test would be conducted on the two levels (combined frequencies) and frequencies for the one remaining level. Summing two levels using the Transform function and Recode into Different Variables option would be the appropriate procedure to combine two levels. However, if data were entered

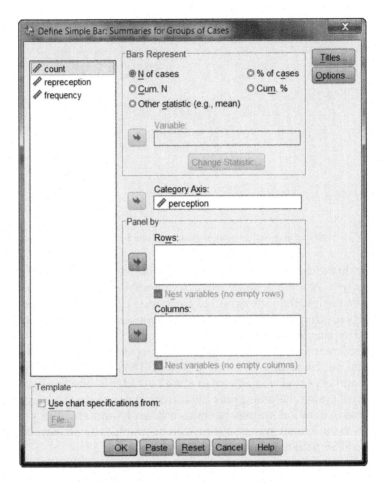

FIGURE 8.25
SPSS dialog box for Define Simple Bar: Summaries for Groups of Cases.

using the weighted cases method, it would be easier to simply sum the two levels to be combined and enter the variable with a new name and new frequencies as shown in Figure 8.27. Unless you change the expected values in the Chi-square Test dialog box, the default will consider that both categories are equal proportions.

The Pearson chi-square test for goodness of fit evaluates differences between observed and hypothesized proportions for levels (categories) of a sample. The proportions are based on the frequencies of responses that fall within each category. The hypothesized proportions may or may not be equal for the categories. Generally, one would consider the proportions to be the same for the different levels unless there is research to the contrary.

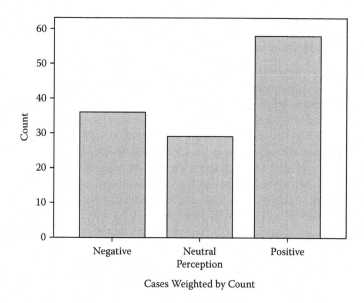

FIGURE 8.26
Frequency of responses for three levels of satisfaction with campus transit system.

FIGURE 8.27
SPSS Data View showing combined cases for 0 and –1.

Recall that in the introduction to this section, another use of the chi-square goodness-of-fit test is to assess the fit of an observed distribution with a hypothesized distribution. When the hypothesized distribution is normal, the data are on at least an interval scale. The Kolmogorov-Smirnov test assumes an underlying continuous distribution, thus making it a better choice than the chi-square test for checking goodness of fit of an observed distribution with a normal distribution. The test can accommodate various kinds of distributions in addition to the normal distribution. The Kolmogorov-Smirnov test is discussed in Section 8.5.

8.5 Kolmogorov-Smirnov One-Sample Test

The Kolmogorov-Smirnov (K-S) test may be used to evaluate whether two sets of data are significantly different from one another. The test compares an empirical distribution (observed sample data) with a hypothetical distribution (expected distribution). Like the chi-square goodness-of-fit test, the purpose of the K-S test is to examine the extent of agreement between the two distributions (observed and unknown). It is highly unlikely that one would know the distribution of the comparison sample. In other words, the test evaluates whether the scores in the sample distribution could have possibly come from a population distribution with the same parameters. This means that different kinds of distributions may be specified for the K-S test. For example, Gaussian (normal), Poisson, exponential, or uniform distributions may be specified. Results of the test are based on the specified distribution. The discussion in this chapter is limited to the normal distribution.

The K-S test for goodness-of-fit compares the cumulative theoretical frequency distribution with the cumulative known (sample) frequency distribution and has several advantages over the chi-square test of goodness of fit. The K-S test is an exact test even for small sample sizes, as it is not limited by minimum expected values as in the chi-square test. It is considered more rigorous and thus more powerful than the chi-square test. Data for the K-S test are assumed to be from a cumulative continuous distribution, making the test useful for ordinal data, whereas the chi-square test is appropriate for nominal data.

The underlying assumptions of the K-S test are that (1) the sample was drawn from a specified theoretical distribution and (2) every observed value is close to the hypothesized value from the theoretical distribution. While some statisticians prefer the K-S statistic to the chi-square, others do not trust the K-S test and recommend that it not be used to test for normality of data. This concern is warranted for the one-sample case because the comparison distribution is seldom identified with known parameters. In such cases, the underlying comparison distribution is assumed generally to be a normal distribution.

The null hypothesis claims no difference between the frequency distribution of a known sample of scores and a specified hypothesized frequency distribution. The null hypothesis is written as follows:

H_0: The data are normally distributed.

The alternative hypothesis is written as follows:

H_A: The data are not normally distributed.

When the null hypothesis is true, the difference between the observed and theoretical distributions will be small and any observed difference will be

$$D = \max |F_o(x_i) - S_n(x_i)|,$$

$$i = 1\ 2,\ 3,\ ...\ N$$

FIGURE 8.28
Formula for Kolmogorov-Smirnov test, where F_o is the hypothesized value for each x_i and S_n is the observed value for each x_i.

due to chance. The K-S procedure tests for the largest difference between the cumulative frequency of the theoretical distribution and the cumulative frequency of the observed sample data. This means that the largest difference is based on the average (cumulative frequency) of the distributions. The value of this largest difference is called the *maximum deviation (D)* as specified in Figure 8.28.

Critical values for the sampling distribution of maximum deviations under the null are known and have been calculated based on sample size. If D is derived through hand calculation, critical values of D are displayed in a table by alpha and N for fewer than 35 samples. For larger samples, the critical value at alpha = 0.05 may be calculated as $D = 1.36/\sqrt{N}$. If the obtained value of D is greater than the tabled value, then one must reject the null hypothesis and conclude that the sample was not taken from the specified population distribution. If the obtained value of D is less than the critical value, one must fail to reject the null hypothesis and conclude that the sample data are from the specified hypothetical distribution.

The test shows the point at which the two distributions show the greatest divergence. The likelihood that such a divergence of the sample data from the theoretical distribution is ascertained based on chance, assuming that the observed sample is a true sample from the theoretical distribution. A random sample from some population distribution is compared with a hypothesized distribution to evaluate whether the random sample represents the true hypothesized distribution. If the test indicates that the two distributions do not agree, one must reject the null hypothesis and conclude that the true theoretical distribution hypothesized in the null hypothesis is not given by the sample.

In addition to conducting the K-S statistical test, special graphical functions such as the cumulative distribution function (cdf) and the empirical distribution function (edf) may be used to estimate the agreement between the random sample data and the specified hypothetical distribution. Graphical functions display the empirical distribution with a dotted line and the hypothetical distribution with a solid line. A quantile-by-quantile (Q-Q) plot or probability–probability (P-P) plot displays a line at a 45-degree angle when the sample data are in agreement with the hypothesized data. It is recommended that one conduct the K-S test, interpret the statistical results, and create the plots. Interpretation of the plots requires experience and sound judgments. Example 8.6 illustrates the use of the K-S test to evaluate normality of a sample distribution.

Example 8.6

The problem for this example is to evaluate whether the number of times a sample of 125 college students, riding a campus transit system for a two-week period follows a normal distribution. A one-way ride counted as one ride; a two-way ride counted as two rides. Students were randomly selected from ten upper-level undergraduate classes. The students agreed to keep records of the number of times they rode the transit system in the two weeks after they were selected for the study. They were instructed to return their anonymous written records in unmarked envelopes provided by the researcher to the faculty member from whose class they were selected. The faculty members returned all envelopes to the researcher. The data were recorded as the number of times the sample of students rode the campus transit for 14 days. The SPSS path to conduct the Kolmogorov-Smirnov test is as follows.

Click Analyze – Click Descriptive Statistics – Click Explore – Move transituse to the Dependent List box – Click Statistics (Descriptives should be selected already as the default) – Click Continue – Select Plots in the Display panel of the Explore dialog box – Click Plots – Select Normality plots with tests – Click Continue – Click OK

Statistical results for the K-S test reveal a *p*-value of 0.17. The non-significant *p*-value suggests that the sample distribution does not differ from the hypothesized normal distribution expected by chance. Therefore, we can conclude that the sample data were drawn from a normal distribution. Results of the test are shown in Figure 8.29. The Q-Q plot for the distribution of scores for the transit use study is displayed in Figure 8.30.

Chapter Summary

The binomial test is appropriate when only two outcomes are possible. In most cases, unless otherwise known, both outcomes are equally likely. A common expression of the binomial test is illustrated by the flip of a fair coin—the outcome of a head is as likely as the outcome of a tail. The binomial distribution converges to the normal distribution as sample sizes exceed 35. Probabilities can be calculated easily with a hand calculator for small ($n = 20$ to 25 trials) sample sizes. The one-sample sign test is concerned with the probability that a random variable will be above or below a preset cut point (usually the median) in the data set. The test uses a plus sign or a minus sign to indicate whether a response falls above or below the cut point. Values equal to the cut point are excluded from the analysis.

If test results are statistically significant, one can conclude that the test variable is different from the cut point; consequently, the null hypothesis is rejected. The runs test produces a test statistic and a two-tailed probability

Tests of Normality

	Kolmogorov-Smirnov[a]			Shapiro-Wilk		
	Statistic	df	Sig.	Statistic	df	Sig.
Transituse	0.073	125	0.168	0.974	125	0.016

[a] Lilliefors Significance Correction

FIGURE 8.29
Results of Kolmogorov-Smirnov test for transit use.

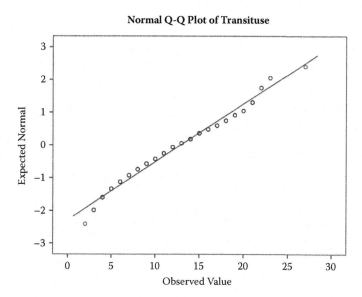

FIGURE 8.30
Q-Q plot for normality of distribution for use of transit data.

that indicate whether a series of values for a variable is random. If the test statistic is significant, one can conclude that the order of the values is non-random and the null hypothesis must be rejected. Conversely, if the test statistic is non-significant, one must retain the null hypothesis and conclude that the series is random.

The Pearson chi-square goodness-of-fit procedure tests whether observed proportions are different from hypothesized proportions for different categories of a sample. The goodness-of-fit test is based on the difference between observed and expected outcomes for each category. The statistic is based on the sum of squares of the differences between observed and expected frequencies divided by the expected frequency. The Kolmogorov-Smirnov (K-S) test compares an observed sample distribution to a theoretical or hypothetical distribution in a population. The hypothetical distribution for comparison is usually the normal distribution. Statistical significance for the K-S suggests that the observed distribution is different from the hypothetical distribution.

Student Exercises

Data files for Exercises 8.5, 8.8, 8.9, 8.10, and 8.11 can be accessed at http://www.crcpress.com/product/isbn/9781466507609, under the Downloads tab in the file Chapter8data.zip.

8.1 What are the various kinds of variables in your field of study that call for dichotomous responses?

8.2 What kinds of situations in your field of study or work may be analyzed with the binomial test?

8.3 Which situations in your field of study or work are appropriate for testing observed outcomes of a task, situation, event, or other occasion with predicted outcomes?

8.4 What are the respective probabilities that a student could guess the correct answer for seven, eight, nine, and ten items when responses on each item are independent of one another and each item has four possible responses? Use SPSS to calculate these probabilities.

8.5 A golf coach at a local community college is preparing the team for an important tournament. The coach has 15 fairly good student golfers, each of whom has played in at least five college tournaments. The average score for the team is 77 on a par 72 course. The coach is interested in the extent to which the team's scores are different from 77 on a par 72 course. Perform the appropriate test to evaluate the null hypothesis: The mean score on a par 72 course is 77 for the ladies' team.

8.6 What are the variables in your area of study or interest that may be tested with the one-sample runs test for randomness? Why would someone in your field of study be interested in the series you selected?

8.7 Construct a data base of 25 values between 1 and 100 using a random number generator. You may use a table of random numbers or computer-generated numbers. Enter the values into the SPSS editor and conduct the one-sample runs test for randomness. What are the results of your analysis? What is your decision about the null hypothesis?

8.8 A teacher of social studies at the secondary school level is interested in the perceptions of her colleagues toward serving students with disabilities. The teacher randomly selected 37 other teachers within her school district to respond to an online 25-item survey. Her return rate is 100%. The teacher is interested in whether her colleagues' responses are random. Conduct the appropriate statistical test to confirm that the responses are random.

a. What is the test value for the teacher's analysis?

b. What is the asymptotic p-value?

c. What is the exact p-value?

d. How many runs did the analysis produce, and what is the probability of obtaining the number of runs revealed in the analysis?

e. What should the teacher conclude about the randomness of the data?

8.9 College students enrolled in teacher preparation programs are required to complete service learning components. The administrator responsible for student placement hypothesizes that there is no significant difference in the proportion of students who want to complete their service in an urban, rural, or suburban school. To test whether students request one type of school over another, the administrator reviews the files of the last 150 placements. If the proportion of students requesting each type of school is the same, the administrator could expect 50 requests in each category.

a. Test the administrator's hypothesis that the proportions of student requests are equal.

b. Are follow-up tests necessary? Why or why not?

c. If follow-up tests are necessary, conduct the necessary follow-up tests.

d. Write the results for this study.

8.10 A human resources officer at a large manufacturing company is concerned about the job satisfaction of the most recent hires. The officer asks a random sample of 60 individuals hired in the last six months whether relationships with co-workers, flexible work schedules, or supervisory support was most important to their job satisfaction. The officer believes that the areas of most importance to job satisfaction are evenly distributed among the 60 new workers. Perform the appropriate analysis and respond to the following items.

a. Test the human resources officer's hypothesis of no difference among the frequency of responses for each of the three areas of job satisfaction.

b. Are follow-up tests necessary? Why or why not?

    c. Should one reject or fail to reject the null hypothesis? Give evidence to support your decision.

    d. Write the results for this study. Include a simple bar graph in your results.

8.11 A science teacher administers a pretest to 141 incoming freshman students. The pretest is scored from 0 to 100. No one is expected to score 0 or even close to 0. The teacher is interested in whether the scores for the incoming students are normally distributed. Her null hypothesis is that the science pretest scores for the incoming freshmen are normally distributed. Test this hypothesis and respond to the following items.

    a. What are the mean and standard deviation of the scores?

    b. What should the science teacher conclude about the normality of the distribution? Give the evidence to support your response.

    c. Create a normality plot to show the distribution of the science pretest scores.

References

Ary, D., Jacobs, L. C., Razavieh, A., and Sorensen, C. K. (2009). *Introduction to Research in Education*, 8th ed. Belmont, CA: Wadsworth Publishing.

Brown, B. M. (1992). On certain bivariate sign tests and medians. *Journal of the American Statistical Association*, 87, 127–135. http://www.jstor.o rg/stable/2290460

Cabana, A. and Cabana, E. M. (1994). Goodness-of-fit and comparison tests of the Kolmogorov-Smirnov type for bivariate populations. *Annals of Statistics*, 22, 1447–1459. http://www.jstor.org/stable/2242233

Dahiya. R. C. (1971). On the Pearson chi-squared goodness-of-fit test statistic. *Biometrika*, 58, 685–686. http://www.js tor.org/stable/2334410

DeGroot, M. H. and Schervish, M. J. (2002). *Probability and Statistics*, 3rd ed. Boston: Addison-Wesley.

Dietz, E. J. (1982). Bivariate nonparametric tests for the three-sample location problem. *Journal of the American Statistical Association*, 77, 163–169. http://www.jstor.org/stable/2287784

Franklin, L. A. (1988). Illustrating hypothesis testing concepts: the student's t versus the sign test. *Mathematics Magazine*, 61, 242–246. http://www.jstor.org/stable/2689360

Gibbons, J. D. & Chakraborti, S. (2010). *Nonparametric Statistical Inference*. Boca Raton, FL: Chapman & Hall/CRC.

Kvam, P. H. and Vidakovic, B. (2007). *Nonparametric Statistics with Applications to Science and Engineering*. Hoboken, NJ: John Wiley & Sons.

Oja, H, and Nyblom, J. (1989). Bivariate sign tests. *Journal of the American Statistical Association*, 84, 249–259. http://www.jstor .org/stable/2289871

O'Reilly, F. J. and Stephens, M. A. (1982). Characterizations and goodness of fit tests. *Journal of the Royal Statistical Society*, 44, 353–360. http://www.jstor.org/stable/2345491

Stephenson, W. R. (1981). A general class of one-sample nonparametric test statistics based on subsamples. *Journal of the American Statistical Association*, 76, 960–966. http://www.jstor.org/stable/2287596

Index